Ray E. Rink

TIME, GARBAGE, GOSPEL

WESTBOW
PRESS®
A DIVISION OF THOMAS NELSON
& ZONDERVAN

Copyright © 2018 Ray E. Rink.

All rights reserved. No part of this book may be used or reproduced by any means, graphic, electronic, or mechanical, including photocopying, recording, taping or by any information storage retrieval system without the written permission of the author except in the case of brief quotations embodied in critical articles and reviews.

Scripture quotations taken from the New English Bible, copyright © Cambridge University Press and Oxford University Press 1961, 1970. All rights reserved.

THE HOLY BIBLE, NEW INTERNATIONAL VERSION®, NIV® Copyright © 1973, 1978, 1984, 2011 by Biblica, Inc.® Used by permission. All rights reserved worldwide.

WestBow Press books may be ordered through booksellers or by contacting:

WestBow Press
A Division of Thomas Nelson & Zondervan
1663 Liberty Drive
Bloomington, IN 47403
www.westbowpress.com
1 (866) 928-1240

Because of the dynamic nature of the Internet, any web addresses or links contained in this book may have changed since publication and may no longer be valid. The views expressed in this work are solely those of the author and do not necessarily reflect the views of the publisher, and the publisher hereby disclaims any responsibility for them.

Any people depicted in stock imagery provided by Getty Images are models, and such images are being used for illustrative purposes only. Certain stock imagery © Getty Images.

ISBN: 978-1-9736-4573-3 (sc)
ISBN: 978-1-9736-4575-7 (hc)
ISBN: 978-1-9736-4574-0 (e)

Library of Congress Control Number: 2018913675

Print information available on the last page.

WestBow Press rev. date: 11/20/2018

Contents

1. **Time: The Mystery** 1

2. **Einstein's Time** 11
   2.1 Special Relativity; Einstein's Train 13
   2.2 Spacetime 24
   2.3 How Big is Spacetime, Anyway? 32
   2.4 General Relativity; Einstein's Turntable 37
   2.5 Black Holes and Time 42
   2.6 Summary 43

3. **Garbage Happens** 45

4. **Time and Entropy** 51
   1. Entropy: What is it, really? 51
   2. Entropy of the Universe 55
   3. Entropy and Irreversibility 56
   4. Time and Entropy 59

5. **Negative Entropy, Creation, and Life** 62
   5.1 In the Beginning, God 62
   5.2 Entropy and The Fall: The Raveling Hole 66
   5.3 The Populous Universe? 67
   5.4 Startrek - Or Not? 69
   5.5 Death and Resurrection 74

6. **The Valley of Bones** 79
   On Memory and Remembering 79

7. **Final Fragments** 84
   7.1 The Compassionate Timekeeper 84
   7.2 Light and Darkness, and Time 86
   7.3 Some Poems about Time, etc. 88
   7.4 The Invitation 96

Author's Preface

By way of introduction, I am a retired engineer and professor of engineering (electrical). But I've also devoted considerable thought over several decades to the mysteries of time, of physics and cosmology, of existence, and of life and death. Having come to a few conclusions, I've felt a need to commit some of these thoughts to paper, in the hope that others might find something of value there. I should here state something about my own preconceptions of ultimate meaning and purpose. I am a Christian – and this fact undoubtedly influences much of my thinking. To be totally up-front about it, my beliefs are summarized in the words of the ancient Nicene Creed.

I am motivated, in part, by the rapidity of advancement in physics, astronomy, and cosmology over the last few decades. Many fundamental questions about the physical universe have been answered in quite satisfactory manner. However, for every answer a host of new questions has arisen, and science seems to have no satisfactory answers to the most fundamental questions of existence, such as "Why is there something instead of nothing?", or "How did the universe begin?" Some writers of popular books seem to bypass the fundamental questions and go directly into speculative accounts of the wonders that await us, with our current knowledge and mastery of technologies such as rocketry, robotics, and artificial intelligence. I am deeply skeptical of such speculations, and hope that the thoughts expressed here may contribute to sober second thought. It is my view that we cannot and will not escape the multitude of societal and environmental problems facing us here on earth by going to Mars or to any other place.

I am motivated also by a concern that many fellow believers seem to have chosen to reject any dialogue with the world of modern scientific research. Those who so choose seem to have made up their minds to the effect that the ancient and traditional cosmology they have inherited is inherently complete and correct, and that there can be nothing new to be learned from science. It is time for believers to rethink those positions in full light of what the other side is saying. We take the Holy Book to be true; we should seek to understand and to assert that truth in the language of our time.

This is not a physics textbook. It does, however, make reference to some topics from modern physics, and does contain a few symbols and equations. I include them in the hope that there may be readers, hopelessly left-brained like myself, who want, *need*, to see some careful, formal development of ideas concerning the physical world. The level of physical reasoning and mathematics is low and can easily be followed by any reader with high-school-level physics and algebra. Those who are intractably allergic to symbols and equations can easily skip past them and read only the wordy parts. Those who find their appetites whetted and want more, much more, can be referred to one of the many excellent textbooks.

Chapter 1
TIME: THE MYSTERY

*Nothing I cared, in the lamb white
days, that time would take me
Up to the swallow thronged loft by the shadow of my hand,
In the moon that is always rising,
Nor that riding to sleep
I should hear him fly with the high fields
And wake to the farm forever fled from the childless land.
Oh as I was young and easy in the mercy of his means,
Time held me green and dying
Though I sang in my chains like the sea.*

From "Fern Hill"[1], by Dylan Thomas

Time. It is a very great mystery, if we think of it. Most people rarely think of it at all, except as a limited resource of which there is never quite enough. The hands of the clock sweep out the minutes of the hour, the hours of the day, and the days of a life, without ever slowing or stopping or reversing. What is past is done and what is to come is unknown. People of faith, who believe that God can influence future outcomes in answer to prayer or to reward the good or punish the evil that men do, generally don't believe that God can change the past. What is past is done and cannot be undone, a profound asymmetry that is called *the arrow of time*. We live, always facing forward, at the cutting edge of history. We may look back, over the shoulder as it were, to remember events of the past, but we can no longer live there and dare not walk backwards into the future.

There are several common ways to think about time, all of them inadequate. There is *objective time*, something which is meted out by our metaphorical clock but is otherwise unknowable. It has no existence as a physical entity; we cannot see, hear, smell or taste it. We can mark down a specific value, such as 13:51:46 Greenwich, but that is useful only for an orderly cataloging of experienced real events. It provides a means of ordering events in sequence and noting the degree of proximity in experience of any two events in the sequence. Objective time as measured by clock and calendar is *linear*, taking on values that are real numbers going arbitrarily far back or far ahead from the present moment. Indeed, one speaks of the time line.

There is *subjective time*, a personal perception of an objective time interval on the part of an individual. This is non-numerical, though, and it is really more a perception of duration, with qualifiers such as short, long, and excruciating. As such, it is highly variable and dependent on many factors such as the individual's state of alertness, wakefulness, emotions, discomfort, etc. An eternity may seem to pass for the husband who is waiting for his wife to complete her shopping, while the day on the golf course passes all too quickly.

Finally, there is *scientific time*, which is entirely abstract and is used as a parameter in the calculus of physical laws that describe dynamical change, i.e. that describe how systems evolve under the influence of cause and effect. It takes on values from the continuum of the real number system and may stop, or progress forward, or reverse, bearing no particular relationship to objective time or subjective time. This kind of time is used in computer simulations of physical processes (e.g. in the exercising of the atmospheric models used by meteorologists to predict weather). If the computer used is powerful enough, many days of real (objective) time may be simulated in a few minutes of computer time, allowing forecasts to be made based on present conditions. The models may also be exercised in reverse time, starting from present conditions, to see if they are sufficiently accurate to be able to account for the atmospheric conditions actually observed at some time in the past.

Physicists, biologists, and theologians have tried to think deeply about some of the questions surrounding the mystery of time. Why must there be an arrow of time? Did time have a beginning? Will it ever end? Where does God stand in relation to time? Evolutionary biologists and cosmologists have tried to make use of what is known of time to explain how life forms and the universe came to be as they are today. But there remain many questions for which no satisfactory answers have yet been found. It is very difficult - some would even say impossible - to think about existence without time. We have no experience of timelessness and can scarcely imagine how it would be. Indeed, thought and imagination are temporal processes. These explorations and their current status will be discussed more thoroughly in subsequent sections of this book.

Philosophers in recent decades have also become very interested in questions of time, especially since Einstein's relativity theories have shown that time is not absolute and may elapse at different rates for different observers, depending on their states of motion. Recent speculation in cosmology has even proposed that many parallel universes exist, each with its own space-time frame that is inaccessible to observers in every other. Many philosophical articles and books have been published, each presenting analysis, arguments, and counter-arguments of some aspect of the metaphysics of time. The philosophical arguments seem mainly to focus on issues of language, semantics, and logic, and are perhaps of less interest to those of us who are not philosophers. For example, one issue having to do with language and time is whether or not *tensed* and *untensed* linguistic descriptions are equivalent. A tensed language is one in which the word *"now"* is used, reflecting a dynamic world view in which only the present is real, while the future is not yet real and the past is no longer real. In this world view *temporal becoming* operates to continuously bring reality into being. An untensed language, by contrast, is one in which the word *"now"* is not used; all events of past, present, and future are equally real and are merely ordered by statements as to the date and time of their occurrence. This usage reflects a static world view wherein the notion of a four-dimensional space-time frame is borrowed from modern physics. In this frame past, present, and future time all coexist along the time dimension. Of course, this fact is of little significance for people who live in the psychological present.

Here I intend to look at how a few poets have sought to express their intuitions about the mystery of time. Poetic sensibility may be as good a tool as any in science to produce at least some useful way of thinking about time beyond the three common ways mentioned above. Poems about time are often filled with expressions of loss and longing, as in the wonderful poem of Dylan Thomas, "Fern Hill", excerpted above. It is indeed inevitable that this be so; those are the natural sentiments that any thinking person must sometimes feel when looking back on a lifetime that is drawing towards a close and cannot be relived. So powerful are these feelings that some (perhaps many) would like to deny the inevitability of the end of life. To this end some biologists are working at understanding the mechanisms of cell aging and death, with the hope of greatly extending human life, perhaps even indefinitely. Should they succeed, this would undoubtedly have profound repercussions for humanity. One can imagine a world filled to capacity with adults who are ancient in calendar years but not in physiological years. They might desire to spend an eternity buying and selling, consuming and excreting, taking pleasure where they can, but having no pressing incentive to soon accomplish anything of much significance. The pressure of limited time would not be acting to spur individuals onward to make the required sacrifices and to do the suffering that is necessary to achieve greatness. It would likely be, for most, a world of gray mediocrity and infinite tedium. Someone would necessarily have to die from time to time, by accident, murder, or execution, to allow for one child to be born. Whatever one believes about the reality or nature of hell, our imagined world would seem to approximate such a place.

The poet Annie Dillard has often touched on metaphysical matters of God and time in her writings. An instance is the following excerpt from "Metaphysical Model With Feathers"[2]

> *This is the shape of the one god, holy,*
> *Who generates the ages, rapt,*

> *Who tolerates time as a hole in his side,*
> *A petrel blind and churning. This*
> *Is the one god, flailed by wings.*
> *And this is the one time, this raveling hole*
> *Swift in god and voiceless, black beak shut.*

A rather mystical theme here is the thought that time is a defect, as it were a wound or raveling hole in the side of God, that has an inevitable course of its own to follow and carries us blindly along. The image is that of a petrel, cruising the breadth of the timeless, sweeping out the ages of reality. If we imagine that time is in some sense a defect, then we might also imagine that the surcease of time could be a healing, a perfection of the imperfect. This idea will be developed further in a later chapter.

The river is another very common metaphor used for time. The phrase *river of time* provides the image of continuous flow, carrying with it all who live in time. An external observer who is outside of time can stand far above the bank of the metaphorical river and view its entire span, both upstream and downstream, much as the physicist can look at 4-dimensional space-time and say that future and past always exist and are not different in kind from present. But the one who is being carried downstream by the current has no such privileged view; he can see only local features as he is swept along.

The idea of continuous flow is misleading, though. We do not actually perceive continuity. What is perceived is a series of discrete images, as the brain processes sensory input, which only seem to merge continuously one into another much as the discrete pictures that make up a movie film merge in our minds to a continuous flow. Also, indeed, we do have some ability to review images from the past through the facility called *memory*.

We also seem to have some ability to subjectively slow down or briefly stop the flow, as when we consciously try to prolong a pleasant moment, being intensely aware of the pleasure for an interval. The following poem brings together these ideas.

"North Saskatchwan Interlude"

The river valley, filled with fog,
Falls away before my feet. Silent
Deep, a hidden world, the river
Bears its deadwood burden down and
Down to the cold and distant Bay
While tangled willows, with the fog,
Help hide the gray and sullen flow
From eyes that do not wish to see.

But up above on valley edge
A bee has found her sun
And hums in ecstasy of warm
Sweet blossoms of the saskatoon.
And there! The red-tail mounts
The shaft of sun-sought air
And soars. Oh god, well done.
For this I live a winter long.

The inexorable flow of the river of time need not always seem sinister, either. A lifetime of memories can hold much that is gratifying. One can feel a deep sense of identity with many of the lives that one has touched over the course of time, and perhaps also with the best accomplishments of one's hands or mind. The poet Wendell Berry has written often on themes of time and mortality and eternity. The following poem[3] expresses beautifully the mixture of regret and peace one can feel upon contemplation of one's life.

> *No, no, there is no going back.*
> *Less and less you are*
> *that possibility you were.*
> *More and more you have become*
> *Those lives and deaths*
> *that have belonged to you.*
> *You have become a sort of grave*
> *containing much that was*
> *and is no more in time, beloved*
> *then, now, and always.*
> *And so you have become a sort of tree*
> *standing over a grave.*
> *Now more than ever you can be*
> *generous toward each day*
> *that comes, young, to disappear*
> *forever, and yet remain*
> *unaging in the mind.*
> *Every day you have less reason*
> *not to give yourself away.*

I have mentioned two linear metaphors for time, namely the clock/calendar timeline and the river of time. A nonlinear alternative to these is the circle or *cycle* of time. This concept may be encountered in Hindu thought, where an individual soul may exist indefinitely through endless cycles of death and reincarnation. In the last century some researchers in cosmology have proposed a cyclic universe, wherein the current expansion phase from the big bang singularity may be followed by a gravitational contraction phase collapsing everything back to a singularity, to be followed by another big bang expansion, and so on through endless cycles. However, current evidence from astronomy does not support this type of cosmology, in that the current expansion of the universe seems to be accelerating rather than exhibiting the gravitational slowing that could lead to eventual contraction. There is also a problem with entropy, a measure of disorder, which seems always to be increasing. There is no known way for the entropy of the universe to systematically decrease, which would be necessary for a cyclic universe.

Nonetheless, many believe that there is always a *renewal* process happening on earth, where life is present. Life itself involves a tendency towards order, a localized decrease in entropy at the expense of greater increase in the entropy of the environment. In Christian thought the Creator is always involved with creation, again and again bringing renewal to bear against the tendency towards chaos. This idea is expressed in the poem "God's Grandeur"[4], by the poet/priest Gerard Manley Hopkins:

The world is charged with the grandeur of God,
  It will flame out, like shining from shook foil;
  It gathers to a greatness, like the ooze of oil
Crushed. Why do men then now not reck his rod?
Generations have trod, have trod, have trod;
  And all is seared with trade; bleared, smeared with toil;
  And wears man's smudge and shares man's smell: the soil
Is bare now, nor can foot feel, being shod.

And for all this, nature is never spent;
  There lives the dearest freshness deep down things;
And though the last lights off the black West went
  Oh, morning, at the brown brink eastward, springs-

Because the Holy Ghost over the bent
World broods with warm breast and with ah! bright wings.

The idea of renewal or re-creation seems slightly different from the concept of the perfect cycle, though, in that it seems to suggest only a sustaining of life (nature, in Hopkins' poem) against the otherwise inexorable decay associated with the entropy-death of the universe.

The spiritual aspects of human existence are inextricably tied up with the experience of time. The apostle Paul's summarizing triad of *faith, hope,* and *charity* are meaningful only if notions of past, present, and future are held. If one held to a static worldview in which the future always exists as fixed reality, this would engender a fatalistic "what will be, will be" attitude that is incompatible with hope. Faith, which has been defined as confidence in - and commitment to - that which is not yet seen, namely the providence of a creator who cares for the world and its creatures, is likewise necessarily connected with the dynamic view of time. Even charity, or love, has its roots in time and experience, as expressed in Wendell Berry's poem cited above.

In the following chapters I shall attempt to outline an unconventional view of the metaphysics of time, one that is consistent with modern physics but also draws on a biblical understanding of reality.

Chapter 2
EINSTEIN'S TIME

Prior to the twentieth century, time was generally treated by scientists as a universal given and as such was simply used as a parameter in verbal accounts of the past, or in graphical descriptions or mathematical models of change. The implicit assumption was that if perfectly accurate clocks had once been synchronized and distributed throughout the universe, they would all forever after show the same time upon subsequent comparisons. This assumption of a universally common and steady flow of time was the basis for much of the scientific work that was done from Galileo onward. The great Sir Isaac Newton, in the late seventeenth century, invented *calculus* and used it to reveal the crucial role of time in his formulations of the fundamental laws of the motions of material objects. He made use of these laws, together with his discovered *law of gravity*, to provide the first satisfactory explanation of the motions of the planets. His assumption about the nature of time is made perfectly clear in his famous statement[5] defining time as "absolute, true and mathematical time, [which] of itself, and from its own nature, flows equably without relation to anything external."

So successful were Newton's laws that they remain in general use today, for calculations of the motions of large (larger than atomic in size) objects moving at speeds much less than the speed of light. From this we may conclude that Newton's assumption about the nature of time is, if not perfectly accurate, adequate for most purposes.

The success of Newtonian mechanics in explaining and predicting planetary motion led many people to assume that all motion, on any scale, could in principle be mathematically determined by such laws and could therefore be predicted far into the future. The term "clockwork universe" has been used to describe such a universe, in which the future is completely determined by present conditions, and the unfolding of that future is as inevitable as the ticking of the universal clock and can only be witnessed, not influenced. The assumptions of such *determinism* have even made their way into the science of human behavioral psychology, as exemplified by the writings of B. F. Skinner.

Twentieth-century physics has, however, broken the yoke of determinism and its associated fatalism. Notions of absolute time and space are gone, replaced by the relativistic time and space of Einstein, and certainty of motion is gone, replaced by the Heisenberg uncertainty principle of quantum mechanics. A great deal of mystery has come back into the modern view of the cosmos, including deep questions such as those regarding the origin and destiny of the universe, the secrets hidden in black holes, the existence of dark matter and dark energy, and the possible use of quantum entanglement as a means of instantaneous communication over vast distances. People for whom the ideas of determinism were spirit-quenching and deadly may now take fresh hope; perhaps all is not as foreordained as previously thought.

2.1 Special Relativity; Einstein's Train

Among the tremendous contributions of Albert Einstein, his 1905 paper "On the Electrodynamics of Moving Bodies" changed forever the prevailing views of space and time. Prior to that paper Newton's view of time, mentioned above, was generally accepted, as was Newton's view of space[6]

> "Absolute space, in its own nature, without relation to anything external, remains always similar and immovable."

Absolute space and time were thought to provide the universal framework within which the physical laws are set. Space was thought by some to be filled with the *ether*, an invisible medium that supports the electromagnetic waves that constitute light, similar to the way that air supports the pressure waves that constitute sound. It was already known, though, from the work of James Clerk Maxwell in the 1860's that electromagnetic waves need no such medium, that they propagate through empty space with a speed that is determined by two parameters, the electrical permittivity and the magnetic permeability of empty space. Looking at a point on the wave, the time-varying magnetic field strength induces the electric field, and the time-varying electric field strength induces the magnetic field. I.e. they support each other.

Einstein's breakthrough came because of an assumption he made, namely

> *The laws of physics are the same in every inertial reference frame, regardless of the velocity with which the frame may be in motion.*

Here "frame' refers to the space/time coordinate system in which the laws are stated, while "inertial" means non-accelerated, moving with any constant velocity. He reported that this thought came to him when, as a teenager, he imagined riding his bicycle along beside a light wave, at the speed of light. He first imagined that the wave would appear to him to be stationary, with *constant* electric and magnetic field strengths, and then realized that this would be impossible, as only time-varying fields can induce one another. Therefore, he realized, the wave would necessarily appear to be moving at the speed of light no matter how fast he pedaled, i.e. no matter what reference frame the observer is in. This can seem paradoxical, if one imagines certain situations. For example, suppose an observer sees a pulse of light from a laser moving past, from left to right, at speed $c=300,000$ kilometers per second, the approximate speed of light in vacuum. Suppose another observer flying past from left to right at speed $c/2$, or 150,000 kilometers per second, later reports also seeing the laser pulse moving from his left to his right at speed c. Observer #1 might be puzzled, as this might seem to imply that the pulse should have been traveling at speed $3c/2$ relative to himself.

Einstein was famous for devising *thought experiments* to gain insight into puzzling phenomena and to intuit possible ways of resolving the puzzle. The thought experiment he used to gain insight in regard to speed-of-light paradoxes is called "Einstein's Train".

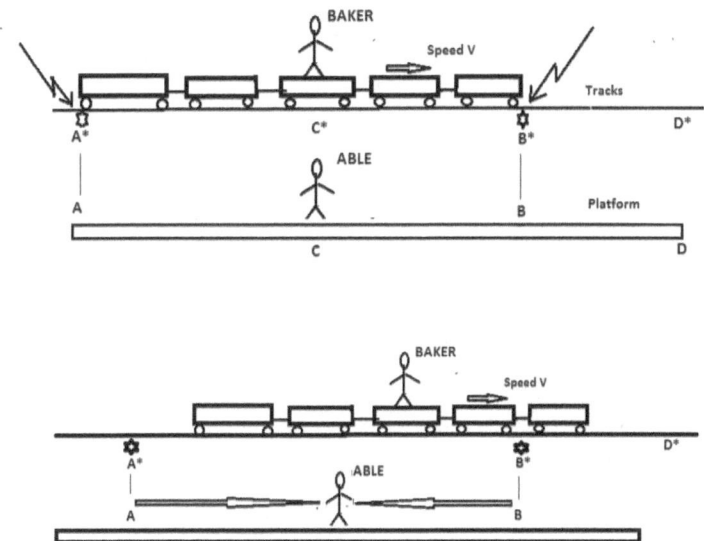

Shown in the upper figure above are twin brothers Able and Baker. Able is the observer on the platform, Baker on the train. Let us suppose that two lightning bolts strike, one striking both the end of the train and the platform at points A* and A, the other striking the other end of the train and the platform at points B* and B. Suppose that Able is positioned precisely midway between A and B, while Baker is positioned precisely midway between A* and B*. Clearly, if the flashes from the lightning strikes reach Able at the same instant (portrayed in second figure above), he would report that the two strikes were simultaneous. He would, however, deduce that the flash from A would reach the current location of Baker only *after* it had passed him (see second figure above), while the flash from B would pass Baker *before* it reached him. He would therefore judge that the two strikes would *not* appear to Baker to be simultaneous. And, indeed, they would not.

Now consider how things appear to Baker. Thus far, we have not used Einstein's breakthrough assumption, that all motion is relative and that the speed of light is c in every inertial frame. Baker can equally well regard his frame (the train) as stationary and the platform as moving at speed V relative to himself. Since the speed of light, coming from each end of the train, is c, and he is at the midpoint, *if* the flashes reach him at the same time he will deduce that the two strikes *are* simultaneous. (But, by reasoning similar to Able's, he would judge that they would not appear simultaneous to Able, and he would be correct in this judgment. Note that this second scenario holds if the lightning bolts struck instead at points A* and D*on the railway track, such that Baker's position in the second figure above is exactly midway between these points.) This demonstrates that judgments of simultaneity are not universally agreed upon when the observers are located at some distance from the events in question and are in relative motion to one another. So the common statement "this event happened here and *at the same time* that event happened there" is often questionable. This is called *the relativity of simultaneity*, and it is the first major departure from Newton's universe.

The second departure has to do with the rates at which time elapses, i.e. the rates of clocks in different frames of reference. (The following discussion is slightly technical, from here to the end of this chapter, but a reader with some knowledge of high-school algebra and a bit of patience will be able to follow it. There is really no alternative, if one wishes to gain insight into Einstein's astonishing contributions. There are many other books which present the results in a non-technical manner, also many which present them at a far *more* technical level. The reader who wishes to avoid all equations can turn to the last section of this chapter for a concise summary of the main conclusions.) Again referring to the first figure above, we will assume that Baker has with him on the train a clock; a very special type of clock called a "light clock". This consists of two horizontal mirrors, one mounted exactly one meter above the other, with a single photon (particle of light) bouncing straight up and straight down between them. (Straight up and down from Baker's perspective, that is, but *not* from Able's perspective.) Baker will note that each "tick" (bounce) of the clock will measure out a unit of time called a light-meter, the time it takes for light to travel one meter at speed c, or $1/c = (1/300,000,000)$ seconds.

From Able's perspective, however, the photon does not move exactly vertically, but rather has a small horizontal component of motion, due to the movement of the train from left to right during the transit. This component has length equal to speed (V, for the train) multiplied by a time Δt, as measured by Able's clock. The distance d traveled by the photon between bounces, according to Able, is therefore found by applying the formula of Pythagorus for right triangles, $d^2 = 1 + (V\Delta t)^2$. Since light travels at speed c in Able's frame, we also have $d^2 = (c\Delta t)^2$ and therefore

$$1 + (V\Delta t)^2 = (c\Delta t)^2,$$

or

$$(\Delta t) = 1/\sqrt{c^2 - V^2} = (1/c)/\sqrt{1 - V^2/c^2}$$

This is slightly greater than the time noted by Baker, and so, if Able could see Baker's clock, he would observe that it is running more slowly than his, by the factor

$$\sqrt{(1-(V/c)^2)}$$

At ordinary train speeds this factor is very close to unity, but if we take the train speed to be, say, 0.6c or 180,000 km/second, the factor becomes $\sqrt{(0.64)} = 0.8$. This is a very significant slowdown for Baker's clock, though he would not be aware of it.

This *time dilation* of clocks in moving frames is the second major departure of Einstein's universe from Newton's. The thoughtful reader will have another question at this point. Since we have merely two frames (train and platform) in relative motion, what if the light clock is on the platform, and Baker is the one seeing the non-vertical path of the photon? Would he not conclude that Able's clock is the one that is running more slowly? Indeed he would, and this demonstrates the fact that the slow clock is the one that is fixed in the frame that is moving with respect to the observer. If there are observers in both frames, they will not agree as to whose clock is slow.

This last observation leads to a famous paradox, the Twin Paradox. To set the stage, let us suppose that the train carrying Baker travels on for one year (in Able's time) at speed 0.6c, then suddenly reverses direction and travels back to Able's platform, arriving there after another year. Although two years have elapsed in Able's life, due to the time dilation effect (which doesn't depend on the direction of motion) only 2(0.8)=1.6 years will have elapsed in Baker's life. So the twins find that Baker is now 0.4 years younger than his twin brother. Strange, but that isn't the paradox. The seeming paradox arises when we consider that, since motion is relative, from Baker's point of view the platform has been on a two-year round-trip journey, and now it returns with Able 0.4 years younger than Baker! So, which twin is younger? They cannot both be.

The answer lies in the fact of the train's drastic deceleration and re-acceleration in the reverse direction. Special relativity does not apply in non-inertial (accelerated) frames, so we are unsure as to Baker's age when he returns. The general relativity theory would be required in order to determine it. Indeed, the relativity of motion doesn't even hold here, since the train and not the platform is undergoing acceleration. The symmetry of the situation is broken.

The third departure from Newton's universe, involving the so-called *length contraction* in a moving reference frame, follows directly from the time dilation in the moving frame. Let us suppose that the platform is 100 meters long, as measured by Able. How long would it appear to be, to Baker? He could note the time shown on his clock when he arrived at the first end of the platform and then again at the second end, subtracting to determine the elapsed time. Able would say the elapsed time is 100/(0.6c)=167 light meters. Baker, however, has a slow-running clock which would show only 0.8 as much elapsed time, or about 133.3 light meters. He also knows his speed to be 0.6c, so he would calculate that the length of the platform to be (speed)(time) = (0.6)(133.3) = 80 meters. I.e. length is contracted in the moving frame by exactly the same factor as time is dilated.

Before moving on, it will be interesting to consider a modification of the Twin Paradox setup, one that will avoid the problem of acceleration and reliably inform us as to Baker's true age. The modification is to suppose that a second track exists beside the platform, and that there is a third brother, Charley, who is traveling from right to left, but still two years (Able's time) away at speed 0.6c, when Baker passes by Able. I.e. not twins, but triplets! Let us also suppose that Baker and Charley are able to synchronize their clocks at the instant they pass each other, one year later (in Able's time) and 0.6 light years to the right of Able. Thus, when Charley goes whizzing by the platform, two years after Baker passes by, his clock will show an elapsed time of only $2(0.8) = 1.6$ years. So he will be 4.8 months younger than Able, as will Baker. The conclusion is unchanged, and there is now no assumed acceleration (never mind how Baker and Charley got to be going so fast in the first place!). Also, there is now no relative symmetry in the situation, by which we could imagine Able as moving instead of Baker and Charley, so there is no paradox.

There is, however, a small question as to the relative speeds of Baker and Charley. If each is moving at 0.6c relative to Able, but in opposite directions, Newton would say that their relative speed is 1.2c. I.e. each would see the other going by at 1.2 times the speed of light. But Einstein would say, not so! The time dilation and length contraction effects prevent this. To show this in general, we need the complete equations of the *Lorentz transformation*, first developed by the Dutch physicist H. A. Lorentz in the late 1800's. This development predates Einstein's, and was done in an attempt to explain why the ether hypothesis did not seem to be verified by experimental observations. However, the equations also follow from Einstein's assumptions.

To develop them, imagine that two coordinate axes (the "x axis" only, since we are considering only one-dimensional motion) are laid parallel to the railway tracks, with time axes in the vertical direction. Denote the first of these space-time coordinate systems (Able's) as x_A, t_A and the second (Baker's) as $x_B, t_B,$ assumed to be moving from left to right with speed V. We want to find the transformation rules that will transform position and time of an event from one coordinate system to another. To find these rules, consider a pulse of light that originates at

$$(x_A, t_A) = (0,0) \text{ and } (x_B, t_B) = (0,0)$$

and travels in the $+x_A$ direction with speed c. At an elapsed time t_A, the pulse will have traveled a distance $x_A = c\, t_A$ in the first coordinate system and a distance $x_B = c\, t_B$ in the second coordinate system, since light has speed c in each frame. Newton would say that $x_B = x_A - Vt_A$ and that $x_A = x_B + Vt_B$. However, knowing of the relativistic contractions of space and time, we let

$$x_B = K(x_A - Vt_A) \text{ and } x_A = K(x_B + Vt_B),$$

where K is some constant parameter, to be determined, that will account for the relativistic contraction of distance (length) in transforming from one frame to the other. Substituting $x_A = c\, t_A$ and $x_B = c\, t_B$ into each of the above two equations, we obtain

$$c\, t_B = K(c-V)\, t_A \text{ and } c\, t_A = K(c+V)\, t_B$$

Eliminating t_A between these two equations gives $t_B = (K^2/c^2)(c-V)(c+V)t_B$, which can be an identity only if

$$K = 1/\sqrt{(1 - V^2/c^2)}$$

which is the same length-contraction factor found previously for Baker's calculation of the length of the platform. From the second of the original two equations we then write

$$t_B = (1/V)[x_A/K - x_B] = 1/V[x_A/K - K(x_A - Vt_A)]$$
$$= K[t_A - (V/c^2)x_A]$$

which, along with

$$x_B = K(x_A - Vt_A),$$

define the Lorentz transformation from coordinates A (Able's) to coordinates B (Baker's).

The equations transforming in the reverse direction can be found to have similar form,

$$t_A = K[t_B + (V/c^2)x_B]$$

and

$$x_A = K(x_B + Vt_B)$$

Now returning to the question regarding the relative speed between Baker and Charley, we can find the transformation between their respective coordinate frames by combining the transformation from Baker to Able, as given above, with an identical one from Able to Charley,

$$t_C = K[t_A + (V/c^2)x_A]$$

and

$$x_C = K(x_A + Vt_A),$$

identical because Charley is moving in the $-x_A$ direction relative to Able, just as Able is moving in the $-x_B$ direction relative to Baker. Now combining the two, by substituting the expressions for t_A and x_A from the first transformation to the second, after some algebraic manipulations we obtain

$$t_c = [t_B + (U/c^2)x_B]/\sqrt{(1 - U^2/c^2)}$$

and

$$x_c = (x_B + Ut_B)/\sqrt{(1 - U^2/c^2)}, \text{ with } U = 2V / (1+V^2/c^2).$$

These expressions have the same appearance as the Lorentz transformation from coordinate system (x_B, t_B) to coordinate system (x_A, t_A), shown previously, except that the relative speed V is replaced, not by 2V as one might expect, but by U. That is, the transformation from Baker's reference frame to Charley's is another Lorentz transformation, but the relative speed of Charley is not 2V= 1.2c but

$$U = 1.2c/(1.36) = 0.89c,$$

which is *less* than the speed of light. This result holds in general, *any* series of Lorentz transformations between reference frames, each moving at a speed less than c with respect to its predecessor, yields an overall transformation with relative speed less than c.

Indeed, according to Einstein's theory, nothing can exceed the speed of light in any inertial frame. We may note that the factor

$$K = 1/\sqrt{(1 - V^2/c^2}$$

which gives both time dilation and length contraction in the moving frame, becomes infinite at V=c. Thus elapsed time is always zero in the frame that is moving at the speed of light. If photons could wear wristwatches, they would not see any time pass on their journeys! A photon from Betelgeuse, 640 light-years away, arrives at your eye as a newborn, freshly emitted from the surface of the star. .

There can be no doubt that the theory of special relativity is correct, as its predictions have been amply confirmed by many experimental observations. Time dilation, for example, has been confirmed by using pairs of initially synchronized, highly accurate atomic clocks, sending one around the world by jet aircraft, and noting that the slight time difference between the clocks upon return agrees exactly with the theory. When we say *correct*, however, we mean it in the sense that any scientific theory is correct, that it accounts for all known observations. That is not to say that no exception can ever be observed. Indeed, a matter of great current interest is that of *quantum entanglement* wherein pairs of separated particles seem to sense one another's quantum state without any elapsed time for data transmission, no matter how distant they may be from one another. This seems to suggest that the speed of light may perhaps not be the absolute speed limit in the universe.

2.2 Spacetime

The results from special relativity, outlined above, may seem profoundly unsettling to some. At relative speeds that are small when compared to the speed of light, the relativistic effects are very small and of no practical significance. Nonetheless, they demolish our intuitive notions that simultaneity, space, and identical because Charley is moving in the $-x_A$ direction relative to Able, just as Able is moving in the $-x_B$ direction relative to Baker. Now combining the two, by substituting the expressions for t_A and x_A from the first transformation to the second, after some algebraic manipulations we obtain

$$t_c = [t_B + (U/c^2)x_B]/\sqrt{(1 - U^2/c^2)}$$

and

$$x_c = (x_B + Ut_B)/\sqrt{(1 - U^2/c^2)}, \text{ with } U = 2V/(1+V^2/c^2).$$

These expressions have the same appearance as the Lorentz transformation from coordinate system (x_B, t_B) to coordinate system (x_A, t_A), shown previously, except that the relative speed V is replaced, not by 2V as one might expect, but by U. That is, the transformation from Baker's reference frame to Charley's is another Lorentz transformation, but the relative speed of Charley is not 2V = 1.2c but

$$U = 1.2c/(1.36) = 0.89c,$$

which is *less* than the speed of light. This result holds in general, *any* series of Lorentz transformations between reference frames, each moving at a speed less than c with respect to its predecessor, yields an overall transformation with relative speed less than c.

Indeed, according to Einstein's theory, nothing can exceed the speed of light in any inertial frame. We may note that the factor

$$K = 1/\sqrt{(1 - V^2/c^2)}$$

which gives both time dilation and length contraction in the moving frame, becomes infinite at V=c. Thus elapsed time is always zero in the frame that is moving at the speed of light. If photons could wear wristwatches, they would not see any time pass on their journeys! A photon from Betelgeuse, 640 light-years away, arrives at your eye as a newborn, freshly emitted from the surface of the star. .

There can be no doubt that the theory of special relativity is correct, as its predictions have been amply confirmed by many experimental observations. Time dilation, for example, has been confirmed by using pairs of initially synchronized, highly accurate atomic clocks, sending one around the world by jet aircraft, and noting that the slight time difference between the clocks upon return agrees exactly with the theory. When we say *correct*, however, we mean it in the sense that any scientific theory is correct, that it accounts for all known observations. That is not to say that no exception can ever be observed. Indeed, a matter of great current interest is that of *quantum entanglement* wherein pairs of separated particles seem to sense one another's quantum state without any elapsed time for data transmission, no matter how distant they may be from one another. This seems to suggest that the speed of light may perhaps not be the absolute speed limit in the universe.

2.2 Spacetime

The results from special relativity, outlined above, may seem profoundly unsettling to some. At relative speeds that are small when compared to the speed of light, the relativistic effects are very small and of no practical significance. Nonetheless, they demolish our intuitive notions that simultaneity, space, and time are the same to everyone, everywhere. Some traditional believers might have thought that the Creator has his dwelling place in space and time, and that space and time to him are the same as they are to all creation. Newton might have thought in such terms, but Einstein could not have done so. Perhaps a more meaningful way of thinking for post-Einsteinian believers is in terms of the "Eternal Now" concept as expressed by post-Einsteinian theologians, namely that the Creator is present in creation not only everywhere but everytime (perhaps it should be called the "Everywhere Here and Eternal Now" concept). He is limited by neither space nor time. The picture that comes to mind is that of the observer who is seated far above the river of time and takes in its entire course from end to end. Included is every point on the river at every time.

Nevertheless, we are confronted by the question: If space and time are so ephemeral, so variable from one observer to another, are space and time even real? That is, do they exist as independent entities, apart from any observer or any matter or energy? Or are they merely abstractions that we use to help classify sensory data? Einstein's famous quote "The only reason for time is so that everything doesn't happen at once" would suggest the latter, though it's not clear that Einstein ever really believed this. A similar reason for space might be "so that everything doesn't happen at the same place".

This question has been one of the most profound issues in the philosophy of physics. The classic Newtonian view that space and time simply *are* is familiar. As such, they are presumed to constitute the stage upon which everything happens, or doesn't happen. For example, the Big Bang explosion and expansion some 14 billion years ago is thought to account for the distribution of all matter and energy in the universe. Newtonians might imagine this as a process of expansion of matter and energy into pre-existing empty space. But others would say no, there is no pre-existence of anything real, that space and time came into existence *with* matter and energy, that, indeed, they exist only in connection with matter and energy. These others would argue that "empty space" is impossible, a meaningless phrase.

This issue is behind one of the most famous paradoxes in physics of the last 350 years, the puzzle known as Newton's Bucket. Think of the following experiment: A bucket, containing water, is suspended from a hook by a rope. The rope is twisted but the bucket is held stationary until it is suddenly released. What happens? At first the bucket starts spinning, due to the torque from the unwinding rope, but the water does not, and its surface remains flat. Soon, however, the water starts to spin as well, due to viscous drag from the wall of the bucket. When this happens, the surface of the water will become concave as water piles up on the bucket wall due to centrifugal force. After a while, the rope will be fully unwound and the bucket will stop spinning. But, due to its momentum, the water will still be spinning, with a concave surface. The puzzling fact here is that we have two periods with relative motion between the bucket and the water; during the first period the surface of the water is flat, while during the second period it is concave. Why? How does one decide that the water is spinning in the second period but not in the first? If the experiment is performed on Earth, with the hook embedded in a beam that is supported by posts set in the ground, there are ample points of reference for that decision. But, Newton argued, what if the bucket is in empty space with no points of reference? Then there would presumably be no basis for deciding whether or not the water is spinning. Newton concluded that space itself must provide the reference, that there must be *something* absolute about space, even empty space, that determines whether or not an object is moving or accelerating.

Ernst Mach, a nineteenth-century Austrian scientist, considered Newton's bucket and postulated that in space that is truly empty there could never be any relative acceleration of the water and hence never a concave surface. But, he argued, if there were distant massive objects such as stars, these would provide the reference frame for acceleration to be observed and hence to cause the behaviour of the water's surface. He did not explain how such distant objects could physically influence the bucket and its contents.

Einstein puzzled over this paradox as well and never fully resolved it, though he apparently leaned towards Newton's argument that there is something absolute about space, despite his insistence that all is relative.

The paradox remains unresolved to this day, though perhaps the way out is to simply insist that "empty space" is a nonsensical concept, that it cannot exist, and therefore to imagine a spinning bucket in otherwise empty space is foolishness. There is no paradox. If space only exists in connection with matter and energy, then these provide the frame in which acceleration can always be defined. For the reader who would like more discussion, Brian Greene's book[7] *The Fabric of the Cosmos: Space, Time, and the Texture of Reality* (Knopf, NY, 2003) presents a highly readable and entertaining account of this and many other issues in modern physics and cosmology.

Returning to the heading of this section, the relativistic nature of space and time as revealed by the special theory has led to the concept of *spacetime*, wherein the frame in which we live, move, and have our being is thought of as having four dimensions, the usual three spatial dimensions (x, y, z) plus time t as the fourth. As a concept, this is nothing new. Good old Newtonian spacetime, wherein all meter sticks are one meter long and all good clocks run at the same speed regardless of which frame of reference is chosen, was the conceptual framework for all physics up to the twentieth century. However, the time and length contractions shown by special relativity for moving frames restrict the usefulness of Newtonian spacetime to only low speeds of motion. A new version was required. This was supplied by Herman Minkowski, a mathematician who was Einstein's high school math teacher (and who, famously, referred to his student Einstein as "a lazy dog"). Minkowski spacetime likewise has three space dimensions (x, y, z), with time as a fourth. It differs from Newtonian spacetime in the *metric* that is introduced to define the distance between any two points. In the Newtonian version the spatial distance d between points (x_1, y_1, z_1) and (x_2, y_2, z_2) is found from

$$d^2 = (x_2 - x_1)^2 + (y_2 - y_1)^2 + (z_2 - z_1)^2$$

and this distance is the same for any observer, regardless of time or motion. In Minkowski spacetime, however, one thinks not primarily of points in space, but rather of *events*. An event is a point in space *and in time*, a primal occurrence that is presumed to happen independently of any reference frame. The metric that defines separation between two events having coordinates (x_1, y_1, z_1, t_1) and (x_2, y_2, z_2, t_2) in a particular 4-dimensional frame of reference is called the *proper time* τ_{12} between these events. It is defined as the square root of

$$(\tau_{12})^2 = (t_2 - t_1)^2 - 1/c^2[(x_2 - x_1)^2 + (y_2 - y_1)^2 + (z_2 - z_1)^2].$$

Consider also a second frame of reference, moving at a speed V relative to the first, with coordinate axes (x', y', z', t'), both frames having common origin (0, 0, 0, 0). Suppose the *same* two events have coordinates (x'_1, y'_1, z'_1, t'_1) and (x'_2, y'_2, z'_2, t'_2) in the moving frame, and define a second proper time between them as the square root of

$$(\tau'_{12})^2 = (t'_2 - t'_1)^2 - 1/c^2[(x'_2 - x'_1)^2 + (y'_2 - y'_1)^2 + (z'_2 - z'_1)^2].$$

Applying the general Lorentz transformation to relate the primed coordinates to the unprimed ones, after considerable algebraic manipulation one can show that $\tau'_{12} = \tau_{12}$, so the proper time is invariant under change from one inertial frame to another. Another name for proper time is *invariant interval*, meaning that it does not vary with a change from one reference frame to another, even to a moving frame that is subject to relativistic clock slowing and length contraction. To summarize, while in the Newtonian frame, we are accustomed to asking "how far is it from point A, where I presently am, to point B, where I want to be, and how long will it take me to get there?", these questions become unanswerable in relativistic spacetime; they need to be replaced by similar questions that have definite answers.

An event in spacetime is said to have two *light cones,* one called the future light cone and the other the past light cone. These are often illustrated with a diagram, like the one sketched below where only two space dimensions x and y are indicated:

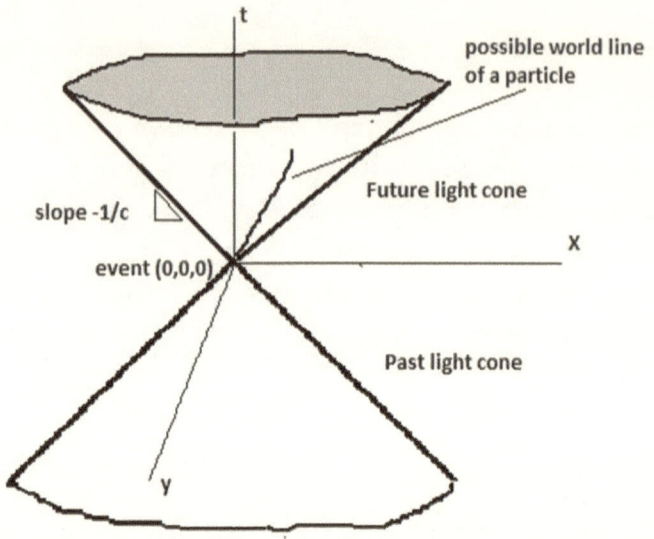

The "world line" of a particle moving into future time from the origin event can only be found within the interior of the future light cone, since it moves at a speed less than the speed of light. An emitted light ray has a world line along the periphery of the future light cone, while an incident light ray arrives along the periphery of the past light cone. The separation in proper time τ_{12} between the origin event and any other event (x, y, t) lying on the periphery of the future light cone is zero, since these events are connected by a light ray and therefore

$$(\tau_{12})^2 = (t - 0)^2 - 1/c^2[(x - 0)^2 + (y - 0)^2] = 0.$$

A world line is said to be *time like* if it lies inside the future light cone (with $(\tau_{12})^2 > 0$), *light like* if it lies on the cone, and *space like* if it lies outside (with $(\tau_{12})^2 < 0$). (In the last case one can define the invariant *space* interval as $-(\tau_{12})^2$, which gives a positive value.) For particles moving on time-like world lines, the proper time τ_{12} is the always the time shown on a clock moving with the particle. It is the same as t_{12} for a particle that is not moving in space, but is less than t_{12} for a moving particle, reflecting the time-contraction of special relativity (t_{12} is Able's time interval, τ_{12} is Baker's).

It is worth noting that the past and future light cones divide spacetime into observable/unobservable and reachable/ unreachable regions. Any past event that lies outside of the *past* light cone is unobservable – it happened at a point in space that is too far away from the observer (sitting at the apex of the cone) for light to have travelled that far in the time available. Any future event that lies outside of the *future* light cone can not be attended by the observer, it will happen at a point in space that is too far away to be reached in the time available. If Able is the observer, and he knows that Baker will have a birthday in six months but that Baker will be 0.667 lightyears away when that happens, he cannot attend. He will be able to see the birthday video eight months later, though, when the birthday event enters his past light cone.

We can interpret the discussion of variations of simultaneity for Able and Baker, from the beginning of this chapter, in terms of events in spacetime. The diagrams of Able's platform and Baker's train are reproduced below for convenience.

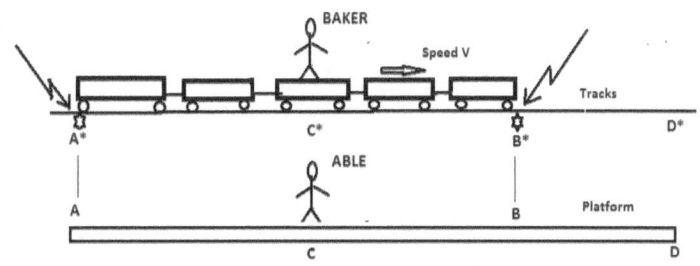

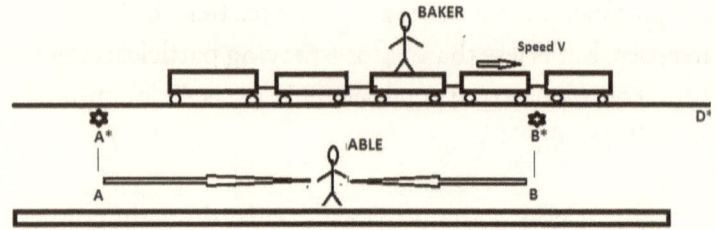

In terms of spacetime, in Able's frame the lightning strikes at events (A*,0), (B*,0), and (D*,0) and their future light cones are as shown below. It is clear that the light cones from A* and B* intersect Able's world line at event P1, so A* and B* are seen as simultaneous by him, whereas the light cones from A* and D* intersect Baker's world line at the event P2 and so A* and D* are seen as simultaneous by him (but not by Able, since the light cones from A* and D* intersect Able's world line at P, and P1 and P are different events).

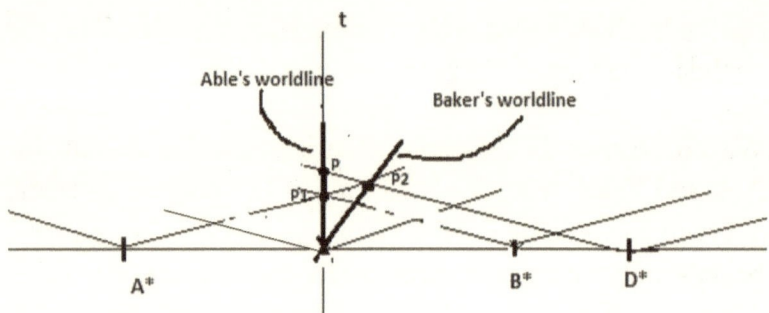

2.3 How Big is Spacetime, Anyway?

In other words, how big is the universe? No one knows for certain. No one knows if it is even finite, rather than infinite. It is generally believed to be unbounded (that is, with no "edges", just as the surface of a sphere has no edges). The widely accepted cosmology at this time is a refinement of the "big bang" theory, wherein the universe came into being by a tremendous eruption of energy (and later on, matter) from a point of singularity, and has been expanding ever since. The current refinement considers the expansion to be a pure expansion of all space, everywhere, rather than a radial expansion from an identifiable central point. The expansion has been far from uniform, though. According to the current *inflation* model, the observable universe expanded from a chaotic primordial region, a tiny nugget that may have been as little as 10^{-26} cm across, and, in an inflation phase that may have lasted only 10^{-35} seconds, expanding in size by a factor of 10^{26}. Since then a slower but smooth expansion, over the estimated 14 billion year life of the universe, has taken it to its present size. We should note that this cosmology describes a possible way for the universe to have evolved – it does *not* explain the pre-existing "tiny nugget", nor is there firm experimental or observational verification of the physics used to explain inflation.

There are two kinds of evidence for this view. One is the observed Hubble *red shift* (similar to the Doppler effect with sound waves, wherin the sound of a train-whistle shifts to a lower pitch when the train passes and is moving away) of the light from distant galaxies, indicating that these are moving away from our galaxy at high rates of speed. The red shift increases with distance, suggesting that the most distant observable objects, extremely powerful quasars, are receding at the greatest speeds (approaching c, the speed of light). The popular analogy suggests that the universe is like a rising loaf of raisin bread, with nearby raisins (galaxies) receding from one another at modest rates. But these rates are compounded as one considers more distant raisins, i.e. raisins twice as far away are receding at nearly twice the nearby rate, and so on. (For very distant galaxies the rate of recession *can* exceed c, the speed of light, unlike the previous discussion of the relative speeds of Baker and Charley. The reason for this is that, with pure space expansion, there is not necessarily any relative speed; the distant galaxy and our Milky Way can each be at rest in its own local space.) The most distant objects seen to date seem to be about 11 to 12 billion light-years away, as estimated from their apparent brightness and energy output. Given this distance, and the rates of recession (red shifts), the estimated age of the *observable* universe is approximately 14 to 15 billion years. The word "observable" is important, as there may be galaxies so far away that any light from them has not had time to reach Earth. If there are such galaxies, and their speed of recession exceeds c, they will never be seen. This is why the universe could even be infinite in size, for all that astronomers know. Unobservable regions may have expanded from primordial regions outside our tiny nugget.

The second kind of evidence for the big bang cosmology is the discovery of the Penzias-Wilson microwave background radiation, low-intensity radio waves that seem to fill the universe and are thought to be the remnants of the extremely intense radiant energy that was once concentrated at the original volume just after inflation. The ratio of the observed wavelength to that assumed for the original energy yields an estimate of the expansion that the universe has undergone, which, by making suitable assumptions, can be made to agree with the estimate of 12 billion light years for the most distant parts of the observable universe.

All such estimates are somewhat questionable, though, depending on several unproven assumptions, assumptions which are frequently revised as new astronomical observations are made. For example, a current hypothesis is that the total mass and energy of the universe may be as much as ten times greater than previously estimated, if "dark matter" and "dark energy" are included. Also, increasingly powerful ground-based and space telescopes are being developed and put into service. If and when more distant objects are observed, estimates of the age and size of the observable universe will increase.

Meanwhile, it is interesting to think about the question "what time is it?" at the far reaches of the universe. If, indeed, objects there are receding from us at very nearly the speed of light, and accelerating as well (see next section), one might think that clocks there are nearly stopped relative to ours, and the time shown on them would be very nearly zero, i.e. Big Bang Time. This is not the case, though, since (just as in the above discussion of relative speeds) with pure space expansion there isn't any common reference frame within which relative speed could be defined, and special relativity does not prevent clocks in every galaxy from showing the same time. There can be a universal time, after all, and perhaps Newton wasn't so far wrong.

To understand this better, we can think a bit more about the Hubble red shift, the down-shift in frequencies of light emitted from distant stars. The frequencies in question are those of the emission line-spectra of the common elements, which are present as well in the distant star. They are emitted from excited atoms in the star as electrons in those atoms fall back to their basic, unexcited energy levels. Therefore they are, or should be, very specific and universal frequencies, common to that element everywhere in the universe. But, for stars that are moving away they are *different*, they exhibit the red shift.

We can understand how this comes about by thinking of a light wave that is travelling through space, from the star to our observer. For convenience, we can imagine that the light is emitted from a continuous-wave (CW) laser that is located on the star. The wave, as emitted, consists of a series of peaks and troughs, successive peaks being separated by a distance (on the star) given by the wavelength,

$$\Gamma = c/fl$$

where c is the speed of light and fl is the laser's frequency. The wave then travels through space at speed c, again as a series of peaks and troughs. But what is the separation between peaks? It is *not* Γ, because, as a given peak moves away from the laser, the laser moves away from the observer by a small distance before the next peak is emitted. In the classical explanation of the Doppler effect, one would say that the given peak moves a distance Γ through space in this time interval, but the total distance between this peak and the next one is the sum of Γ plus the small increase in distance between the laser and the observer. This sum is then the wavelength seen by the observer, hence the red shift. Meanwhile the laser, a kind of clock, is still operating at frequency fl, so this clock has not been slowed at all by the recession speed of the distant star. (The Hubble red shift is not exactly the Doppler effect, though, in that the emitter is not simply moving away from the observer in a fixed frame of reference. What is happening is that *all* of the distance between emitter and observer is being stretched by the expansion of space, so the "small distance" by which the laser is displaced is actually spread over the entire path from star to Earth. The effect on received wavelength is still the same, though, as is the conclusion about the frequency of the laser.)

However, light from the most distant visible objects has needed almost all of the 14-billion-year life of the universe to make the journey to Earth, so the light we see from these objects could be light from the creation, from the Let-There-Be-Light moment. Perhaps it is not surprising that such objects (Quasars, quasi-stellar objects of unknown structure) appear to radiate astonishing amounts of energy.

2.4 General Relativity; Einstein's Turntable

All of the above discussion of special relativity was premised on the assumption of NO ACCELERATION, all frames of reference and all objects within those frames were assumed to be either stationary or moving at constant speed relative to one another. We can now turn to a case where there is acceleration, and ask what happens to clocks and meter-sticks there. Einstein had another brilliant thought in this regard. In thinking about gravity, he realized that a man in free-fall, in a gravitational field, does not feel his own weight and, if he is inside an elevator with no external references, cannot know whether he is falling or is floating in space, far from any gravity field. If he is falling, he is accelerating, so the sensations of weight and acceleration must cancel each other. Einstein deduced from this his famous *principle of equivalence*, which states that gravity and acceleration are entirely equivalent. He also deduced that a light-beam, crossing the free-falling elevator from one wall to the opposite wall, as seen by an external, stationary observer would have to be curved - otherwise the elevator's occupant would have a clue that he was in fact accelerating and not floating in outer space. Since light, by the laws of physics in Newtonian space, can only travel in straight lines, Einstein deduced that *space itself must be curved* in the presence of a gravitational field, or in the presence of acceleration. From this observation, after 10 years of hard work and some help from outstanding mathematicians, he was able to develop the mathematical description of how the geometry of space is altered by the presence of mass and its consequent gravity.

We shall not go there. Instead, let's consider a simple thought-experiment that Einstein used to illustrate the effect of acceleration on space and time. It will suffice to show that the effect of acceleration on clocks and meter-sticks is similar to the effect of velocity. The imagined situation is that of a man, let's call him Able, who is sitting at the center of a large turntable that is spinning very fast. We assume first that Able is seated on the stationary axle and is not rotating, but is gazing outward in a fixed direction at the perimeter of the turntable as it goes whizzing past him. (Here we must assume that the experiment is taking place in a given *inertial frame*, as established perhaps by the "fixed stars" of the Milky Way galaxy, within which absolute rotation can be defined. Recall the need for this as discussed previously in connection with Newton's Bucket.)

Let's also assume that clocks and meter-sticks are distributed all along the circumference of the turntable, and that Able can see them and record their rates and lengths. What will the record show? If clocks and sticks are moving past at very high speed, nearly the speed of light, Able will see the time-contraction and length-contraction effects of special relativity, just as he did for Baker's train. Namely, the clocks will be running slow relative to his timepiece and the meter-sticks will be too short. Indeed, if the rotational speed of the turntable could be increased such that the consequent velocity of a perimeter point reached the speed of light, the clocks would stop and the sticks would vanish! (This cannot be done - the energy required would be infinite.)

Next let's assume that Able is no longer seated on the axle, but is instead seated on the platform's hub, so that he rotates with the turntable. Now, if he gazes outward along a fixed radial, he will focus on only *one* perimeter clock and *one* meter-stick. What will he record now? Recalling that the contraction effects are real, not mere illusion, Able will again note that the clock is running slow and the stick has shrunk. But now he will be puzzled as to the reason for this, if he knows only about special relativity, because he sees no relative velocity. However, if he takes a step or two outward, along the radial, he will experience a very strong centrifugal force, pulling him towards the perimeter. If he resists this force, he experiences the strong centripetal acceleration produced by the resisting force. At this point he may say "Aha! It is acceleration that is causing the clock to slow and the stick to shrink!" And that, in a nutshell, is it.

There arises a question of geometry, however. If the meter-sticks around the perimeter of the turntable all are shortened, the circumference C must be shortened as well. But the diameter D is not (only lengths that are parallel to the direction of motion are contracted), so what has happened to the formula that says C=(pi)D? The surprising answer is that the disc of the turntable is no longer flat, but is cupped such that the circumference of its rim is less than (pi)D. This illustrates the curving of space when acceleration or mass (gravity) are involved. But Able will not notice this effect, since the light ray that enters his eye is also curved and appears to come in along the original plane of the turntable.

We can try to go one step further and relate the acceleration in this setup to gravitational attraction. Suppose now that the turntable is not rotating but that, instead, items on its perimeter are attracted inward by the presence of an extremely massive point object located at the center hub. We can ask: What amount of mass would be required to cause a clock at the perimeter appear to slow markedly, or even to stop? We can consider Newton's Law of Gravity, which states that the acceleration produced by a mass M at a distance R is

$$A1 = GM/R^2$$

where $G = 6.674 \times 10^{-11}$ meters per second per second. For the case of the rotating turntable, if a point on the periphery is moving at speed V the centripetal acceleration is

$$A2 = V^2/R.$$

Equating these two expressions, we find that

$$M = V^2R/G$$

would be the mass required to cause gravitational acceleration equal to the acceleration that is due to rotation of the turntable.

This expression is reasonably correct, for modest velocities V. It is not correct, however, for very high velocities that approach c, the speed of light. For V=c, the correct expression is

$$M = c^2R/G/2$$

i.e. only half as great. The reason for this error is that Newton's Law of Gravity breaks down at such speeds and accelerations;

to obtain the correct result one would need to solve Einstein's gravitational field equations. (We are in good company, however; in one of his early papers on general relativity Einstein predicted the angle through which starlight passing near the sun would be curved - incorrectly, by a factor of 2.)

The field equations were solved by Karl Schwartzchild in 1916, for the simple case of a point mass in otherwise empty space. His solution predicts that every point mass has a Schwartzchild Radius, given by solving the above equation for R, namely

$$R = 2MG/c^2$$

This expression gives a radius at which the gravitational field becomes so strong that clocks, as seen by an external observer such as Able, appear to stop. It is a small number for modest masses; even for a mass equal to that of Earth (about 6×10^{24} kg) the Schwartzchild radius is only about 9 mm. Of course, that mass would have to be concentrated into a spherical volume of radius less than 9 mm for the expression to be applicable. Ordinary bodies such as planets and stars are all much larger than their Schwartzchild radii and therefore do not stop clocks. Even a neutron star, the densest body in the known universe, typically has a radius greater than its Schwartzchild radius.

However, many cosmologists and astronomers believe that massive objects called "black holes" exist in the universe, perhaps at the centers of galaxies and elsewhere, which are denser than neutron stars and have radii smaller than their Schwartzchild radii. No light can be emitted from such an object because, at the Schwartzchild radius where clocks appear to stop, all frequencies of radiation appear to go to zero. Hence the descriptor "black". These objects, it is thought, may be formed when a very large star (having mass more than 10 times that of our sun) explodes as a supernova, losing some of its mass and energy, with the remnants undergoing "Big Crunch" gravitational collapse to an unknown state of matter occupying a region of extreme density, perhaps even a singularity like that once thought to have been at the origin of the Big Bang..

2.5 Black Holes and Time

If Baker could approach a black hole and Able could observe that Baker's clock slows towards a full stop (as Baker appears also to do) as Baker approaches a certain radius, the Schwartzchild radius, what does Baker observe? The situation is similar to that of Einstein's train, where Able sees that Baker's light clock is running slowly, but Baker is not aware of this. So, likewise here, Baker observes his clock to be running normally as he plunges towards and through the Schwartzchild radius. Indeed, since Baker is in free fall he is not aware that he is being accelerated towards a very sudden end on a very massive object. (At some point, though, he will sense, painfully, that his body is being severely stretched, due to gravity strength variation along its length.) So, unlike the train situation where Baker rides off into the sunset, aging more slowly than his twin, this situation has no happy ending. Able may think that Baker is hovering at the Schwartzchild radius, remaining forever young, but actually Baker is long (!) gone.

Such time distortions in the presence of strong gravitational fields have led some theoreticians to speculate about the possibility of time travel, wherein time could not only be slowed but actually reversed. The decade of the 1990's was marked by considerable efforts to find mass distributions that might produce very special spacetime curvatures, ones that could support trajectories having closed time loops, along which one could presumably travel, moving first backward and then forward in time. (Think *Startrek,* time warps, and wormholes.) Although not forbidden by relativity theory, these efforts have by now been largely abandoned, both for practical and philosophical reasons. Practical: One estimate of the energy needed to keep open a one-meter-wide wormhole is 10 billion times the energy production rate of the sun. Philosophical: Time travel to the past would violate the principle of causality, in that a present action could alter a past outcome, which could prevent the present action, … Think of the old paradox about a time traveler going back in time and preventing his parents from meeting, thereby preventing himself from existing, thereby preventing the time travel …

2.6 Summary

This chapter has presented a brief description of Einstein's theories of special and general relativity. The purpose in this is to give the reader an introduction to the modern concepts of spacetime, the 4-dimensional framework that has replaced Newton's concepts of absolute 3-dimensional Euclidean space and absolute time.

The theory of special relativity concludes that, for objects in motion relative to a stationary observer, clocks are slowed and lengths in the direction of motion are contracted by the factor

$$\sqrt{(1 - V^2/c^2)}$$

where V is the speed of motion and c is the speed of light in any non-accelerated frame of reference. This factor is very close to unity for all speeds significantly less than c, which is to say for all speeds encountered in common human experience. Only subatomic particles or, conceivably, spacecraft on very long voyages to other planets or stars are, or might be, accelerated to relativistic speeds where time and distance may be significantly altered. Nevertheless, the fact that time *can* be slowed represents a chink in the armor of the notion of uniform aging of everything and everyone that exists. It also casts into doubt notions of simultaneity and judgments of "before" and "after". However, another way of measuring the separation in time between events in *spacetime*, called the *invariant interval*, removes these ambiguities.

The theory of general relativity extends this conclusion to situations where extremely strong gravitation fields or extremely high accelerations may occur, perhaps in close proximity to the hypothetical "black holes" thought to possibly exist in the universe. General relativity also shows that mass and gravity cause space to be curved, completing the demolition of Newton's ideas that time is regular and space is everywhere flat.

Following chapters will present discussion of the related idea of *entropy*, which is a measure of the degree to which a system is unstructured or disordered, or of the extent of decay of an originally-ordered system. It will be argued that the general trend for overall entropy to always be increasing is related to inexorable increase in age, of which clock or calendar time is merely a surrogate. One might guess, then, that the end of time would mark the end of death and decay, and, perhaps, the restoration of perfect order.

Chapter 3
GARBAGE HAPPENS

> *Turning and turning in the widening gyre*
> *The falcon cannot hear the falconer;*
> *Things fall apart; the centre cannot hold;*
> *Mere anarchy is loosed upon the world,*
> *The blood-dimmed tide is loosed, and everywhere*
> *The ceremony of innocence is drowned;*
> *The best lack all conviction, while the worst*
> *Are full of passionate intensity.*

From "The Second Coming", by William Butler Yeats[8]

We live in a universe that is slowly, but surely, falling apart and growing cold:

1. The universe is expanding, galaxies are moving away from one another at ever-increasing speeds. The galaxies themselves are expanding, separations between stars are increasing.

2. The supply of hydrogen that fuels stars is gradually being consumed. Existing stars must die as they exhaust their fuel supply, and the rate of formation of new stars from free hydrogen must eventually decrease.

3. Within our solar system, the sun will cool and become a "red giant", expanding and destroying the inner planets, including earth. Meanwhile, due to tidal forces, Earth is slowing in its rate of rotation and revolution about the sun, so days and years are gradually lengthening.

All of the above outcomes will, thankfully, happen only very slowly, over time spans of many billions of years. However, when we narrow our view to the here and now, there exist other lists of dire projections that may come to pass much sooner. Fifty years ago it was feared that the "population bomb" would explode and Earth would not be able to support its population, leading to mass starvation or wars of annihilation. That fear has subsided, somewhat, as birth rates in most countries have fallen. More recently, environmental concerns have become critical; some point to increasing concentrations of carbon dioxide in the atmosphere, due to increasing combustion of fossil fuels, with the prediction of runaway global warming or, at least, some degree of unspecified climate change. Environmental degradation has already reached alarming extremes; an example is the floating continent of plastic waste, twice the size of Texas, in the Pacific Ocean between California and Hawaii.

Also of great concern is the proliferation of nuclear-warhead-tipped missiles in arsenals around the world, with fears of a world-wide nuclear holocaust (the Doomsday Clock has recently been advanced to 2 minutes, 30 seconds 'till midnight). Others point to the run-away urbanization trend of recent decades, with the corresponding loss of survival skills and increased vulnerability to catastrophes of all kinds. Artificial intelligence, robotics, and automation threaten to take away many present jobs, leaving most people unemployed and impoverished while owners of the automated industries become ever wealthier.

And the list goes on. Governments and central bankers have learned that they can print unlimited quantities of money and carry unlimited amounts of national debt with no thought of ever repaying any of it. Crypto-currencies have emerged as alternative forms of money, with no limit on the number of them. Hyper-inflation would seem to be the only possible consequence. Need we mention crime, opioids, device addiction, family breakdown?

All of these dangers, and many more, lead some people to dream of escape, perhaps to another planet. The only candidate suggested as of this date appears to be Mars. Within the last five years, seemingly serious proposals have been made to launch a first, one-way expedition to Mars, with a few brave would-be settlers who would seek to establish a self sustaining colony there. This despite the fact that Mars has no breathable atmosphere, little or no available soil or fresh water, and mean surface temperatures on the order of -80 degrees Fahrenheit. Elon Musk has proposed that the first Mars expedition could launch as early as 2020, using some of his rockets. Many thousands of people around the world have volunteered to be pioneers.

This is madness. It would be far easier to establish a self-sustaining colony at the South Pole, where at least there is breathable air, plenty of water, and adequate sunlight and air temperature for about 4 months of each year. But that would not capture the public imagination, whereas an expedition to Mars would occupy the television news networks 24/7 for as long as there was even a single survivor.

For people of faith, however, the most ominous degradation of all is the great spiritual extinction that has been taking place for the past 200 years or so, particularly in the West. (Francis Schaeffer has written at length on the decline of the Christian Church and the effects on humanity that this is having, in "How Should We Then Live? The Rise and Decline of Western Thought and Culture"[9], and in other writings.) The historian Yuval Noah Harari has recently published two bestselling books, "Sapiens: A Brief History of Humankind"[10] and "Homo Deus: A Brief History of Tomorrow"[11]. The latter book, in particular, should make for chilling reading for any thoughtful member of a faith community.

Harari asserts, following Nietzsche, that the great monotheistic religions of the last three millennia are now dead, having been supplanted during the past 200 years by the religion of our time, evolutionary humanism. Every religion centers around worship - so who or what do humanists worship? They worship humanity, Homo Sapiens, who as the object of worship can now be known as Homo Deus. As such, Homo is now, according to Harari, on the threshold of immortality. He asserts that medical science now has the knowledge and the technology to maintain the life of an individual forever, or at least is on the threshold of such knowledge. True, some organs or other body parts may fail, but these can be replaced by artificial devices, producing *cyborgs*, beings with a mixture of organic and manufactured parts. What is more, with eternal life Homo Deus cyborgs can explore the solar system or the galaxy at their leisure. So what if the destination star is one thousand or one million light years away? Homo Deus can go there and thus can become truly the masters of the universe.

Harari suggests that quality of life for such beings (us) can be whatever is desired, since with unlimited computing power and brain imaging all of consciousness can be mapped and manipulated. Happiness? No problem! Indeed, functions of the brain itself may be all or in part taken over by AI, artificial intelligence.

And so on. The book reads like so many bad science fiction novels of the past, but it purports to be serious, non-fictional. This reader is unconvinced. As to the claim for eternal life, one remembers a declaration made by Richard Nixon, in 1971, of war against cancer. The claim was that, with sufficient expenditures, cancer would soon be defeated. That was nearly fifty years ago, and today, after the expenditure of hundreds of billions of dollars in cancer research, most metastatic human cancers are still incurable. As for AI, those who design computers know that a computer, despite the power of its CPU, or the speed and size of its memory banks, or the cleverness of its program, is merely a glorified adding machine. True inventive genius requires something more.

At the root of the Homo Deus proposition is an assumption - that a living human person is only a machine and, as such, needs only an AI mechanic to make her thrive forever. Denied are the existence of good and evil, of the soul, of inspiration, and of transcendent mystery. Einstein, arguably the greatest scientific mind who ever lived, would not have made such assumptions. His task, as he saw it, was to try to understand something of what the Old One had created.

Denied also by the proponents of evolutionary humanism is any higher source for moral and ethical social behavior. Ever since Jean-Jacques Rousseau we have heard the maxim "I need only to consult myself with regard to what I wish to do; what I feel to be good is good, what I feel to be bad is bad". The repeated first person pronoun, from the name of God, I Am, is appropriate for Homo Deus; now if only HD could attain the eternal presence of the verb Am, he might be onto something. Not.

In this brief chapter I have tried to mention some of the many ways in which our world seems to be degrading. True, there have been times when light has come into the world. Chapter 1 of the Gospel of John tells of such a time and gives us hope.

Chapter 4
TIME AND ENTROPY

Entropy: A measure of unavailable energy: a measure of the energy in a system or process that is unavailable to do work.

1. Entropy: What is it, really?

It is evident from the above definition that the notion of *entropy* comes from the field of energy engineering, called *thermodynamics*. While everyone has some notion of energy, the concept of entropy is slightly less familiar and more subtle. A simple example may help to illustrate the difference. Consider the setup shown below, where a hot body at temperature T_1 and a cold body at temperature T_2 are connected through a *working body*. The working body may be simply a heat-conducting element, or it may be some type of heat engine that can convert heat to work.

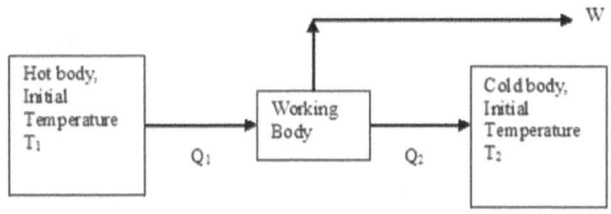

TIME, GARBAGE, GOSPEL | 51

In the diagram, Q_1 and Q_2 are quantities of heat energy taken from the hot body and delivered to the cold body, respectively, while W is the quantity of work done. The law of conservation of energy, also called the *first law of thermodynamics*, states that energy cannot be created or destroyed; it can only be converted from one form to another, for example from heat energy to mechanical work. For the above setup, if the working body loses no heat, this law takes the form

$$W = Q_1 - Q_2$$

Entropy is mathematically defined as $S = Q/T$, heat divided by temperature. In the setup, if $W=0$, then $Q_1 = Q_2 = Q$, and the entropy change associated with the exchange of heat is

$$\Delta S = Q_2/T_2 - Q_1/T_1 = Q[1/T_2 - 1/T_1],$$

and ΔS is positive since T_1 is greater than T_2. The *second law of thermodynamics* states that the entropy change associated with any physical process in a closed system is non-negative. It can be zero, for a reversible process, but is always positive for a non-reversible process. In other words, entropy can be created but never destroyed.

For the setup, this implies that heat will always flow naturally from the hotter body to the colder, never in the reverse direction, and so the process is irreversible. Of course, heat can be *made* to flow in the reverse direction if W is negative, i.e. the heat engine is a refrigerator. But then the system is not closed; the work W is coming in from outside the setup, and the second law does not apply.

Getting back to the original definition, we can see that, for the setup, the heat flow causes the temperature difference to decrease, reducing the amount of work that can still be done. As the temperature difference approaches zero, no further heat flow is possible and no further work can be done. Then the system has reached equilibrium and, even though there is still heat energy in the two equal-temperature bodies, none of it is available to do work.

Another form of the second law is the statement that *there can be no perpetual-motion machine*. This law has not stopped crackpot inventors, from time immemorial, from trying to invent such machines. (Even Leonardo, definitely not a crackpot, tried his hand at it.) Patent offices around the world have received many submissions concerning such would-be inventions. Truth be told, the second law, like all laws of physics, is not really a law of the kind that can be formally proved to be true – it's more a matter of universally-observed fact. It is not impossible that someday a violation of this "law" will be observed, though no physicist expects that to happen.

The concept of entropy is not limited to thermal processes, but, with suitable definition, can be extended to any physical process. Just as in the above setup, where an *ordered* state (two bodies at *different* temperatures, low entropy) naturally relaxes to a *nonordered* state (all bodies at the same temperature, high entropy), for general processes entropy is a measure of the degree of disorder. The greater the disorder, the higher will be the entropy value. For a more complex system, such as a box full of gas particles, disorder is related to the distribution of particles and their velocities within the box. If all were found in one corner, moving in the same direction, this would be a highly ordered (and extremely unusual) state of affairs. That is to say, it would be a state that has a very low probability of occurrence. After a very short time, due to collisions with the walls of the box and with each other, the particles would, with high probability, be found to be quite uniformly distributed throughout the box and with randomized velocities. In other words, the distribution relaxes and the entropy increases very rapidly to its "normal" (high) value for a highly disordered state.

The Boltzmann formula for the entropy associated with the ith state of such a system is

$$S_i = k \log(p_i)$$

where k is Boltzmann's constant, 1.38 x 10^{-23} Joules/degree kelvin, "log" is the natural logarithm, and p_i is the probability of that state occurring. Since probabilities are always less than unity, and the logarithm of such a number is negative, approaching negative infinity as the number goes towards zero, S_i has a large negative value for improbable states and a small negative value for highly probable states (since only differences in entropy are important, one can add a very large positive number to all entropy values, making them all positive). The entropy of the system, at a point in time, is found by averaging over all states,

$$S = \sum p_i S_i .$$

The Second Law says that entropy of a closed system never decreases with time, implying that distributions always tend to relax towards their most probable, disordered states.

2. Entropy of the Universe

Unlikely as it might seem, theoretical physicists have attempted to calculate the entropy of the universe and to speculate as to what implications the Second Law might have for the future of said universe (one of the major contributions of the late Stephen Hawking was to find a formula for the entropy of a black hole). As might be expected, because the observable universe contains vast numbers of galaxies, stars, and other objects, plus interstellar gas, the estimates of the entropy of the present universe are staggeringly large numbers; one such estimate[12] is

$$S_U = 10^{123} \text{ Joules per degree kelvin,}$$

that is, a 1 followed by 123 zeros. This is estimated[12] to be 10^{35} times greater than the Big Bang entropy.

Of greater interest are the speculations as to the past and future values of S_U, beginning with the big bang and going out to whatever the end of the universe might be. One assumes that the initial entropy was relatively low, with all of the energy contained in the very small region known as the big-bang singularity (similar to all of the gas molecules being found in one corner of the box). *If the universe is considered to be a closed system,* then the Second Law would seem to imply that entropy has been increasing ever since, and that the disorder and chaos can only increase as the universe continues to expand and cool. But this is not a pleasing prospect for most Homo Deus humanists to accept. Evolutionary humanism believes in progress. Our domain is imagined to be getting slightly better every day in every way, despite the evidences to the contrary mentioned in the last chapter.. A long, slow slide to oblivion? No, thank you!

In order to escape the nesessity of this conclusion, cosmologists have been trying to find mechanisms by which the entropy of the observed universe might, at some point, begin to decrease and the universe stop expanding and become more ordered. One such possibility might be the existance of heretofore unknown reservoirs of low-entropy orderliness that can dilute the known high entropy.

Another way out is to drop the assumption that the universe is a closed system. Looking back at the simple setup diagramed at the beginning of this chapter, if "work" comes in from outside the Second Law is disarmed and entropy of the open system can decrease.

Most physicists are loath to take this way out – if the physical universe comprises everything that is; then how can there be any "outside"? (Some of the more adventurous ones speculate on the possibilities of "parallel" anti-matter universes – but these, if they exist, cannot come into contact with ours. Also, there is no evidence for the existence of other universes.)

3. Entropy and Irreversibility

It seems to be a regrettable fact of life, as we know it, that all real physical processes are irreversible. Oh, if a book topples and falls off from a shelf, someone can pick it up and put it back, but this as *not* a reversal of the original falling process. The "someone" has to enter the picture and give a hand. Books do not, in human experience, jump back from the floor to the shelf by themselves. This is despite the fact that the laws of physics, which describe such things as falling books, *are* reversible.

How so? Gravity causes the falling book to accelerate downward, towards the floor, gaining kinetic energy as it falls. When it strikes the floor, with a thud, most of that kinetic energy is transferred to vibrational energy in the floor and to sound waves in the air, while some has been transferred to velocity of air molecules that were pushed aside during the fall. The laws that govern such things as vibrations and sound waves do *not* forbid the possibility that, somehow, the spreading vibration and sound wave patterns could be exactly reversed later on (perhaps reflected from the walls of the library?) and, converging precisely together could jar the book upward with the exact velocity needed to cause it to rise to its perch on the shelf. On its return journey, the disturbed air molecules would have to do their part as well, returning to give back the kinetic energy that they gained during the fall – otherwise the book would not have quite enough energy to rise all the way to the shelf.

Might one expect to ever witness such a reversal? Certainly not! After the Fall, the library system has become more disordered – the potential energy of the book, nicely perched on its shelf, has been distributed to other, diffuse kinetic energies of vibration and random particle motion, No sane person would hope that the original orderliness would spontaneously be restored. The Second Law says that the entropy of the library has increased, and that it cannot now spontaneously decrease to its original value, despite that the laws of mechanics do not forbid this.

Brian Greene, in his book *The Fabric of the Cosmos*[7], provides a nice discussion of a similar (but even messier) irreversible event. This time it is a raw egg that is balanced on the very edge of a kitchen table, which, because of some slight disturbance, rolls off and smashes on the floor, egg shell fragments and contents splattering in every direction. Again, by the same arguments as above, the laws of physics do not prevent this disaster from spontaneously reversing itself. Every minute particle of splattered egg *could*, miraculously, bounce back together in such a way as to fuse together and form a perfect egg, the egg with just the right upward velocity to return to its precarious balance on the table edge.

Again, no sane person would expect to see that happen. Were it to happen, it would truly be regarded as a miracle. But wait – there *is* a way it could almost happen. What if the kitchen also contained a hen, a hungry hen that could go round and peck up every fragment of the shattered egg? Thus nourished, the hen could then easily perch on the table and, after a suitable interval, lay *another* egg, identical in almost every respect to the original. Were the second egg to fall, the same renewal process could repeat itself. It is not an exact path reversal, of course, the egg falls by itself but rises and reappears only with much clucking and fluttering of chicken wings.

Has the Second Law been defeated? Well, not quite. The problem is that the hen, while extremely efficient, is not perfectly efficient. She will need a small amount of additional sustenance in order to keep functioning in this manner. This additional amount would be an *input* to the kitchen system – thus the system would not be a closed system and the Second Law would not apply.

But this near-miracle does illustrate a very important fact; namely that living organisms have the ability to reverse local increases in entropy. The hen is able to reverse the increased entropy of the shattered egg, a local increase, at the expense of only a small input from ouside the kitchen. Because of this small input, the entropy of the universe has increased during the overall process, but the local entropy of the kitchen has been restored.

One might be tempted to *define* a living organism as one that has the capacity to reduce its local (internal) entropy and thereby sustain itself. Such a definition would mean, however, that the thermal setup shown at the beginning of this chapter might be regarded as "living", if the working body is an automatically-controlled refrigerator and comes "on" as needed to keep the temperature of the cool body low. Most people would not want to regard an automatic system of this type as "living", though there are some who might do so if pressed. The present confusion surrounding artificial intelligence and robotics is centered on such inappropriate definitions. Even a robot can keep the ice cream frozen, but that signifies nothing.

4. Time and Entropy

In the first chapter I mentioned several kinds of time, with the comment that all are inadequate to encompass our experience of life. There was *clock time*; what is shown on a timepiece at a particular instance and location in Newtonian space; *subjective time*, one's perception of duration; and *scientific time*, in Einstein/Minkowski spacetime. Here I shall propose a fourth kind of time, one that seems more likely to correspond with life process as we experience it.

This is based on the observation that entropy, like time, has an "arrow", a direction that never reverses. By the Second Law, entropy of a closed system can only increase, or remain constant, with time, it cannot decrease. If one were to sketch a plot of this entropy versus time, it would be a curve that never goes down as one looks further out in the first quadrant. If the axes of the plot are interchanged, one has a curve of time versus entropy. This curve might have segments that are vertical straight lines, where time is increasing but entropy is not, but if one takes a new definition of time as "entropic time"

$$t_E = S$$

each sucessive value of S corresponds to greater value of t_E and the curve of entropic time versus entropy is a straight line at a 45-degree angle in the first quadrant. Thus *entropic time* always and only increases with increasing entropy, that is to say with increasing disorder within the system. As such, it would be more revealing, than is clock time, of the processes of aging and of the remaining life expectancy of a system. Note that the mysterious "arrow of time" is now explained by the "arrow of entropy", i.e. by the Second Law.

It is often noted that some people "age" more rapidly than others. Some seem to be aged already at 55, others still relatively youthful at 75. The process of aging could be more closely related to entropic time than to clock time. But how could entropic age be measured, if that were desired? (Life insurance companies might well like such a measure.) Perhaps by the length of the telomere sequences at the ends of chromosomes? Such information, even if available, might not be welcomed by most people. Though, when one is playing the game of life, it may be good to know when the game is in the fourth quarter, especially if one is behind on the scoreboard. The entropic age of one's car might also be valuable to know, as a clue that it may be time to trade it in on a new one.

Some obvious problems with entropic time are 1) It is difficult to define and measure, and 2) It elapses at different rates for every system – there is no universal timepiece that shows "what entropic time it is". Ordinary clock time, while not very useful for gauging the life expectancy of a car, or of a person, or of a universe, at least is easy to define and to read, and certainly is useful for making and keeping appointments. It is not a measure of the entropy of anything, except in the case of a windup clock where it measures the entropy of its spring.

Chapter 5

NEGATIVE ENTROPY, CREATION, AND LIFE

If you do not believe in God
Go on a blue spring day across these fields
Listen to the orchids, race the sea, scent the wind.

Come back and tell me it was all an accident
A collision of blind chance
In the empty hugeness of space

Kenneth C Steven, "Prayer", from "Iona, Poems"[13]

5.1 In the Beginning, God

The disconcerting problem with the ever-increasing entropy of the universe, mentioned in the previous chapter, raises some fundamental questions that seem unanswerable with today's knowledge of physics and cosmology. Among these are:

1. How did the universe, and time and space, begin?

2. From whence did energy and matter come?
3. What was present at the beginning to give an extremely high degree of orderliness (extremely low entropy) to the nascent universe?
4. What end is in store for the universe? Will it indeed continue forever to cool and expand and become increasingly disorganized, appoaching a final state that is a cold, thin "particle soup"?
5. How could life have begun? Did it begin elsewhere than on earth?
6. How should humans see their role in the greater scheme of things?

Certainly some theoreticians have been thinking about such questions and have been trying to formulate some possible answers. Readers who would like to follow those lines of thought will find many readable accounts. (Brian Greene's book[7] would be a good place to start.) In my opinion, these questions, and any possible answers, are, ultimately, outside the realm of the physical sciences. Of course, everyone has a right to think about such matters and, indeed, *should* be thinking about them.

People of faith, however, will have a rather unique and different perspective when it comes to ultimate questions of being. To begin with, we will *not* make the assumption that our physical universe is a closed system. We may well believe that there is another realm that is outside, in some sense, or, if not ouside, is somehow beyond the reach of scientific observation. We might call this the spiritual realm, or, God.

If the observable, physical universe is not a closed system, it is an *open* system; open in some way to an influence that comes from elsewhere. If so, the Second Law does not apply – negative entropy can come in, and the entropy death of the universe is not certain. Indeed, people of faith understand that such influences can help to explain the otherwise-utter mysteries of questions (1) to (4) above.

For believing Jews and Christians, explanations start and end with the Holy Book, the Bible. There it is revealed to us that (Genesis 1:1-5)[14]

"In the beginning God created the heavens and the earth. Now the earth was formless and empty, darkness was over the surface of the deep, and the Spirit of God was hovering over the waters. And God said 'Let there be light', and there was light. God saw that the light was good, and he separated the light from the darkness ... And there was evening, and there was morning – the first day."

Already in this brief passage from Genesis 1 we find some answers to questions (1) to (3) above. To question (3) the answer is: God. To question (2): God's Word. To question (1): God's Will.

To be sure, these are not the kind of answers that theoretical physicists are looking for (but not finding). They look for answers based on a physics of causality, cause and effect, incompatible with questions of First Causes. For instance, the Big Bang theory, as advanced by some, has the Universe beginning at a single point, the Initial Singularity. But others have argued that it cannot be a point – a point is a mathematical abstraction, and physics doesn't happen at a point – so the initial space must have preexisted as something slightly larger, at least a few Planck lengths (1.6×10^{-35} m) in span, and already the game is lost. The inflationary big-bang model mentioned in Chapter 2 starts with a small region, not a singularity, but gives no account of how that region came to be and to have concentrated within it all the energy of the universe.

Genesis 1 goes on to describe the remaining five days of creation, culminating in the creation of Homo Sapiens. Among believers, people of faith, there has been some disagreement as to how this was, and how long it took. The interpretation of the word "day" in particular, has been controversial for some. Most, of the "long day" school of thought, would regard the word as referring to an epoch, an unspecified but probably long period of time, as in the usage "In my day, we always wore suits". Others, of the "short day" persuasion, insist that it refers to the time taken for one rotation of earth about its axis. I would prefer not to take a position on such issues, but cannot resist noting that the earth apparently did not rotate on its axis until Day Four. That cat is probably out of the bag, anyway; in previous chapters I have not disputed the currently-accepted numbers for the age of the universe.

The other great debate among Creationists is with regard to what roles, if any, geological processes and biological evolution played in the process. Again, to this point I have not expressed doubt concerning the geological record; so to claim neutrality here would be disingenuous. True, God *could* have created the Grand Canyon as it now is, to fool us into thinking that erosion did it, but why would He? I've heard no good answer for that. The God that I trust in doesn't try to fool people. As to biological evolution, small evolutionary changes seem to be observed in the present era, so how (and why) should it be totally denied? In Genesis 1 we read repeatedly phrases such as "*Let* the water under the sky be gathered to one place, and *let* dry ground appear", and "*Let* the land produce vegetation", and "*Let* the water teem with living creatures". The permissive construction suggests the intelligent design and approval of natural processes, rather than outright hands-on intervention. This is not to say that all living organisms evolved from a bit of primordial slime that was struck by lightning, or some such event. If that did happen, it would anyway constitute a miracle of God-like proportions, one that would put Brian Greene's Second-Law-defying restoration of the shattered egg to shame. Proponents of evolution-only argue that it *could* have happened, given some organic chemicals to start with and some 3 billion years of elapsed time and millions of minute changes. That undirected, accidental change could have given us a Mozart and an Einstein, well, no, sorry, I am not that credulous.

When we turn to Day Six, the creation of Homo Sapiens, we find something different. The reading here is "Let us *make* man in our image, in our likeness, and let them rule over the fish of the sea and the birds of the air..."; no longer passive/permissive, but, rather, active. Something very different seems to be at work here – HS is not just another link like every other in the biological chain. Indeed, anthropologists suggest that something very unique and inexplicable happened about 70,000 years ago[10] in the prehistory of our species. It is called the Cognitive Revolution, referring to the rather sudden appearance of new ways of thinking and communicating, notably the first use of abstract, categorical language, something that no other species has achieved. Homo became Homo Sapiens, the Wise Ones. Day Six?

5.2 Entropy and The Fall: The Raveling Hole

Here I'll come back to a promise made in Chapter1 in connection with Anne Dillard's poem.

In the account of Genesis 1 we see repeated the phrase "and God saw that it was good" at the end of each day of creation. Indeed it was, not only good but wonderfully perfect.

But it didn't last. With Homo Sapiens came consciousness, and self-consciousness, and will. Opportunity for willful disobedience arose, and "Self" came into conflict with harmony. That which had been flawless became flawed with sin, guilt, and fear. The "raveling hole in the side of god" in Dillard's poem is time, entropic time, with all its implications of growing disorder, death, and decay. I believe that general healing can come only on That Day when time is no more. That Day will come, not only for the children of God but also for the entire universe:

> *For the creation was subjected to frustration, not by its own choice, but by the will of the one who subjected it, in hope that the creation itself will be liberated from its bondage to decay and brought into the glorious freedom of the children of God*[15]

When that happens, I believe, the Universe will cease to be an open system but will become closed, God no longer transcendent and beyond, but now making Earth his dwelling-place.

5.3 The Populous Universe?

People of today seem to want to believe that life must exist in many places in the universe. *We are not alone* is perhaps a comforting thought, once we have denied the possibility of transcendence and of a God who cares. The arguments in favor of extra-terrestrial life are usually based on some kind of probability statement. The informal argument often goes something like this:

> *There are known to be trillions of stars out there. Astronomers have discovered that many of them have orbiting planets, so there must be billions of planets. Even if only one in a thousand of those are within the "life zone" distance from its star, not too hot nor too cold, there must be millions of such special earth-like planets. Since we know that life evolved here on earth, it it highly probable that life came to exist in many other places as well. How could it be otherwise?*

The difficulty with such a statement is that it invokes a false notion of probability. Probability theory is a branch of mathematics that deals with formally-defined events in probability spaces. Probabilities of these events are defined, as givens. In reality, with real events, probabilities must be *estimated*, as *statistics*, from sample data. The size of the sampled-data-base must be large enough so that the required estimates can be made with confidence.

In the present case, we have a sample size of ONE; we know about only one planet, Earth, and it is a planet that harbors life. An extrapolation from that fact to a statement of the probability of life on any other planet is impossible and, therefore, probability theory cannot be properly used. Proponents of life-elsewhere belief are left with only a sort of intuitive argument, that, given billions of years of existence and billions of light-years of space, how could it be possible that life has *not* evolved in many places? Implicit in such arguments is the idea that if one simply waits long enough and has enough room, something astonishingly good *must* eventually happen. Not only must it happen, but also it must become the generating event of an entirely different future, for instance a universe that teems with intelligent life.

Such thinking flies in the face of the stark reality of the Second Law, which implies something more akin to Murphy's Law, namely

> If Something Bad Can Happen, It Will.

In a universe that teems with randomly-moving, energetic particles, while the occasional collision might lead to formation of a more complex molecule, it is extraordinarily unlikely that any such local fluctuation would be a branching point that leads to a sustained chain of development within which the universal tendency for increasing disorder is forever reversed.

5.4 Startrek - Or Not?

Believers do not know if life exists anywhere else in the universe. Some, knowing how special and precious is Earth to the Creator, might be inclined to doubt that it does. But most would simply take a neutral position on the matter. Unless life is found to exist elsewhere in our solar system, or an unambiguous radio message is heard on one of the SETI receivers, we will likely never know. The nearest stars with planets that might be candidates for finding life are more than 10 light years away, a 100-year trip at one-tenth the speed of light. This is likely greater than the life expectancy than any star-ship crew member, so a voyage to search for life or to colonize another star system would seem infeasible.

But wait, *only* one-tenth the speed of light, $V = 0.1c$, or 30,000km per second? That may not seem too fast (and couldn't one go a lot faster?), until one does the math. To accelerate a starship to that speed involves giving it a kinetic energy $0.5MV^2 = 0.005Mc^2$, M being its mass. From Einstein's formula $E = Mc^2$ relating mass to energy, one sees that a mass of about one-half of one percent of the ship's final mass would have to be converted to kinetic energy to achieve that final speed. If the ship were made of pure uranium, enriched to reactor-grade fuel, and if *all* of that uranium could be made to undergo fission, nuclear physics tells us that about one-tenth of one percent of its mass would be converted to thermal energy. This is only one-fifth of the kinetic energy needed, and therefore the maximum speed achieved would be less than 0.1c/sqrt(5). Much less, in fact, since the engineering reality is that much of the thermal energy is lost in the process of converting it to the thrust that is needed to accelerate the ship, and the one-way trip to search for life would take much longer than 100sqrt(5) years. If the energy-conversion efficiency is 50%, the duration of a ten-light-year trip would be 100sqrt(10) = 316.22 years.

The picture is less bleak if one realizes that only the *final* mass needs to reach the maximum speed. Spent fuel can be ejected (rearward, at high velocity, to provide the reaction mass and momentum that a rocket engine needs to produce thrust), and a large percentage of the initial mass can thus be shed as the fuel is consumed during the acceleration phase. If, for example, the final mass is only one-tenth of the initial mass, with constant thrust the final acceleration is ten times greater than the initial acceleration, and an easy calculation shows that the final speed is six times greater than predicted in the previous paragraph. So, a 10 light-year, one-way trip might be accomplished in about 55 years, depending on how much of that time can be spent at maximum velocity. This is a considerable improvement; the concept is similar to that achieved through use of multistage, liquid- or solid-fueled booster rockets as presently used to put heavy payloads into low-Earth orbit.

Which raises another question. Upon approach to the destination star system, the starship would need to be *decelerated* in order to achieve orbit near some planet. This again requires energy, firing rocket engines to apply thrust in the reverse direction. Not so much energy, however, since the approaching ship has only one-tenth of its original mass. Nevertheless, an amount of nuclear fuel equal to nine-tenths of its remaining mass would be required to achieve the needed deceleration. So, in total, 99% of the original mass that departed Earth orbit would need to be fuel. If the colonist's payload is, say, 100 tonnes, the initial mass of the starship plus fuel, leaving Earth orbit, would be 10,000 tonnes. And, note, this leaves no fuel at all for a return trip!

Could better results be obtained if fusion, instead of fission, power could be used? If the D-D (deuterium) reaction is used, about 225 terajoules of energy is released per kg of deuterium consumed. Since the fuel would likely be in the form of heavy water, D_2O, with 4/20 of its mass being deuterium, about 45 terajoules of energy is released per kg of fuel (the oxygen portion would also be useful, to provide reaction mass for the rocket engines, since hydrogen is so light). However, to bring the spacecraft to the speed 0.1c requires imparting to each kilogram of final mass a kinetic energy $0.5(9 \times 10^{14})$ or 450 terajoules. If the final mass is now 10% of initial mass, a final speed 0.6c/sqrt(10) could be achieved if the energy conversion were 100% efficient. At 50 % efficiency, only half as much useful energy is available, so the final speed could be 0.6c/sqrt(20) and the one-way trip takes about 75 years. This is not as fast as the trip with fission

(55 years), and moreover controlled fusion power has not yet been achieved, even on Earth, so this is still doubtful. What is more, the size and mass of a deuterium-burning controlled-fusion power plant, if ever achieved, may well prove to be an insurmountable obstacle for use in starship propulsion.

Some visionaries engage in blithe speculation about interstellar or even intergalactic travel at speeds approaching light-speed, thereby achieving relativistic time-contraction so that occupants do not die from old age prior to arrival, but they offer no practical ideas about how to achieve such speeds. (In the above calculations, the greatest speed achievable was just under $0.2c$, at which speed clocks would be running at about 98% the rate of Earth-based clocks, so the starship occupants age by only about one year less than their earth-based cohorts during a 55-year voyage.) Some have proposed wildly radical propulsion schemes for achieving interstellar travel. One such, which I call the Hiroshima Propulsion Scheme, involves detonating a series of small fission or fusion bombs in the wake of the spacecraft, in the hope that the radiation pressure from the blasts would eventually impart enough speed to the spacecraft! This is a modification of the multistage booster idea, and I would deduce that it cannot compete, even on paper, with the controlled-reaction schemes mentioned above, since the nuclear material must be carried along with the starship and the efficiency of conversion from reaction energy to kinetic energy must be much lower for a bomb blast than it is for a controlled and contained reaction. Also, the practical considerations stagger the mind.

It is also, perhaps, conceivable that physics research will reveal entirely new sources of energy that fundamentally change the conclusions reached above. After all, 100 years ago no one could imagine the energy concentrations now available through nuclear reactions. We cannot, therefore, be certain that human travel to distant parts of our galaxy or even to other galaxies will *never* be possible, though it is highly unlikely. A further, enormous complication with speeds approaching the speed of light is the relativistic increase in mass. This was not mentioned in Chapter 2, but it is in fact true that the mass of an object that is accelerated from rest increases by the same factor $1/\sqrt{1-V^2/c^2}$ that governs time and length contractions. The mass becomes enormous as V approaches c, becoming infinite at V=c. Even at 0.6c, the mass of the starship is increased by 25% above its rest-mass value. All of the energy in the universe would not suffice to accelerate a starship from rest to the speed of light.

Some realists, recognizing that human travel to the stars will not be possible in the foreseeable future, take a different tack. They suggest that robots with advanced AI technology *can* survive a journey of hundreds or thousands of years and can, in a sense, "populate" our galaxy and others. This idea is similar to Harari's ideas[11] about immortal cyborgs carrying out the Homo Deus agenda. Others have suggested that *constructors,* autonomous machines designed to mine the needed minerals and to build self-sustaining bases on distant planets, could serve the robot or cyborg settlers. As an engineer, I am highly skeptical. Perhaps, if we can ever perfect driverless cars and pilotless airplanes that do not crash ...

One can only wonder about the fascination people have with space travel, and about the motivations of political and scientific leaders who promote these ideas. Reaching for the stars is a romantic idea, perhaps especially so when conditions here on earth are quite desperate. When so many in this world are starving and homeless, and when political leaders are haunted by visions of failure, it may be tempting to try and divert the attention of the people to outer space. Perhaps it is the modern variation on the "bread and circuses" stratagems of ancient Rome. It must fail, as it did for Rome.

5.5 Death and Resurrection

This is the topic that one hesitates to think or to write about, if only because death is the great unknown and the ultimate weapon in the arsenal of the Enemy. Some would suggest otherwise; that death is our friend because it is, at last, release from a painful life. It may be that, but only if life is painful. It was not meant to be so. The proper stance against death is found, I think, in lines of the poet Dylan Thomas[16]

> *Do not go gentle into that good night,*
> *Old age should burn and rave at close of day;*
> *Rage, rage against the dying of the light.*

This is not to say that we are helpless and hopeless in the face of death. We are not, because we have the promise of deliverance, the good news that came in the person of Jesus Christ. He demonstrated, in his life, that he has power over death. He promised that those who trust in him shall have access to that power and shall, like him, not perish forever, but shall be with him in the life to come.

This we believe, but what that life will be is still unclear. We are creatures caught in the web of time; we cannot imagine how escape will be. To be sure, many have tried to imagine *future* places called heaven, or hell, and have speculated how they might be. It seems that often such speculations derive from the conditions of life that the speculator has experienced. For instance, people who have had to toil far too hard in life may like to imagine a heaven that has no work to be done, only leisure enjoyment. Others who have been impoverished through life might imagine heaven as a place of neither hunger nor any other want; only great riches, streets paved with gold, etc.

There is a little story that, I think, illustrates how we may be misled by our imaginings. It is an old story (author unknown) about a man who always had far too much work to do. In addition to his demanding job, he always seemed to have more work around his home than he could possibly get done. There was always grass to mow, flower beds to weed, windows to wash, fences to paint, roof to mend, car maintenance, etc., etc. But one morning he awoke and, much to his amazement, could find nothing at all that needed doing. It was a beautiful, sunny day, and very quiet - not even any dogs barking or traffic in the street. As he sat in his living room and thought about the day ahead, he began to feel slightly bored and wondered just what he could do with this day. Finally he noticed the mailman, walking down the sidewalk towards his house. Relieved, he decided to go out and greet the mailman, but as he did so he noticed that the mailman's bag seemed to be empty, with no mail to deliver. He said "Hello, friend, isn't this a beautiful day? I see that you don't have any work to do either. I guess we must have died and gone to heaven!" The mailman seemed surprised at this greeting and, after a long pause, said "Oh. You must be new here. This isn't heaven, it's hell."

Such a concept of hell certainly differs from the medieval image of a place of everlasting, burning fire and demonic torment as punishment for the wicked. Scholars tell us that the association with fire comes from an Old Testament word for hell, *Gehenna*, which referred to the burning trash heap that was located in the valley of the same name that was located outside the Jerusalem city walls. The notion of a trash heap certainly seems appropriate for some of the evildoers who have lived among us, quite apart from the fire aspect. One thought about hell, and only one, seems certain to me. That is that no one is sent there against their will – some willingly go there because that is where they want to be. They want to be as far away as possible from the Lord of life and from those who love him. (Not an original thought, no less an authority than C. S. Lewis[17] has written that "the gates of hell are locked from the inside".) They are given leave to go there, if that has been their choice, as in the Lord's final words of dismissal to them "depart from me, I never knew you"[18].

If hell is quite unimaginable, so is heaven. I can only believe that it will be a great gathering of all the souls who have lived in hope *"that the creation itself will be liberated from its bondage to decay and brought into the glorious freedom of the children of God."*[15]. In this multitude will be all whom we have known and loved and who have loved us. It is incumbent on us in the here and now to speak the truth about this, to extend the Gospel invitation to all. Where will this gathering be? Why, here, on (the renewed) Earth, God's creation, as described in the vision of St. John[19],

Then I saw a new heaven and a new earth, for the first heaven and the first earth had passed away, and there was no longer any sea. I saw the Holy City, the new Jerusalem, coming down out of heaven from God, prepared as a bride beautifully dressed for her husband. And I heard a loud voice from the throne saying,

"Now the dwelling of God is with men, and he will live with them. They will be his people, and God himself will be with them and be their God. He will wipe every tear from their eyes. There will be no more death or mourning or crying or pain, for the old order of things has passed away."

And when will this take place? Impossible to say. Some would say that soon (in cosmological time) a great cosmic catastrophe will destroy the first earth and John's vision will then be made manifest. Others would say not so, it may be billions of years before the present universe will pass away in the course of its entropic death. In the latter case there would seem to be a very long wait in the grave, before the Resurrection. But no, personal time stops at the point of death. When the eyes close in death, *at the very next instant* they open again and it is Resurrection morning. No waiting. To be sure, this question has occupied the minds of theologians. Some have argued that heaven exists now, in parallel with the present earth, and that those who have died in Christ are there now, alive, waiting for us to join them. One piece of evidence for this comes from the words of Jesus to the repentant thief on the nearby cross[20] *"today* you will be with me in paradise". But, others counter, this "today" may be taken as referring to the dying men's "today", the time that it is for them at the point of death and forever after. These would say that believers who have died are, until the Resurrection, in the condition called "soul sleep". They are sleeping, not conscious.

Many of the biblical references to this question are found in the letters of the Apostle Paul. I Corinthians, chapter 15, is the great chapter on the resurrection. It seems to me that the belief in "the resurrection of the dead", as stated in the Nicene Creed, has no meaning if the dead are not really asleep until that day.

There are no scientific answers to theological questions such as those raised in this section, in the sense that at least some questions in physics can be answered. Anyone interested in the debate will find many volumes, some much larger than this book, on the arguments and counterarguments for the various theological positions. We can only trust that the biblical prophecies and promises are revealed truth and that this sad universe remains open to the restoring influence of its creator. On the personal level, we confess our ignorance and pray for deliverance in words such as those from the old hymn[21]

> *Swift to its close ebbs out life's little day;*
> *Earth's joys grow dim, its glories pass away;*
> *Change and decay in all around I see;*
> *O thou who changest not, abide with me.*

Chapter 6
THE VALLEY OF BONES

On Memory and Remembering

In Chapter 2 it was stated that time travel *backward* in time is impossible, because it violates causality. (In contrast, time travel *forward in time* is easy, we do it all the time at the rate of one second per second. It can even be speeded up, if we can accelerate ourselves to something approaching the speed of light. Riding on Einstein's train, at speed 0.6c, Baker can return to visit his brother Able at a time that is 2 years in Able's future but only 1.6 years in Baker's future.)

There is a kind of backwards time travel that is possible, though. It is accomplished by the exercise of human memory. We can, in a sense, visit the past, but we cannot change anything there and cannot cause the present circumstances to be changed. But we can allow memories to influence our present actions and thus to change our future from what it would otherwise have been. This is an amazing gift, when we think of it; it really is a kind of time travel. Our memories largely make us to be who we are, and enable us to do what we do.

Human memory is still quite poorly understood, in general. Short-term memory is thought to be relatively simple; an ongoing pattern of electrochemical brain activity that keeps a piece of information active and current for as long as it persists. Long-term memory, however, must be related to structural changes in neural connectivity that are established through memorization or repeated sensory experience. It is further subdivided into *semantic* memory, concerning abstract knowledge of principles or facts, and *episodic* or *autobiographical* memory, concerning personal experiences. Researchers generally believe that the structural change corresponding to a specific long-term memory is not localized; rather it involves a distributed pattern of connectivity in the brain that is somehow able to be "triggered" when the memory is retrieved, producing a wave of neural activity that takes us back, in a sense, to the fact or experience that we wish to retrieve.

The retrieval of memories is a mystery. How do we trigger the wave that produces the memory, unless we already remember what we are trying to remember? It's a chicken-and-egg problem. Perhaps there must always be some other stimulus, of which we may not be consciously aware, that does the triggering.

However, I'm more interested in the significance of memory than in the neurophysiology. It seems that the amazing gift we call "autobiographical memory" is unique to our species. Other species certainly have ability to learn and to store information about their environment, and to adapt and make appropriate responses to stimuli. It seems, though, that they do not spend time "remembering" without explicit stimuli. They live always in the present. I noticed a convincing example of this in connection with a family dog named Mink that we had many years ago. We planned to be overseas for a year on a study leave, and needed to leave Mink behind. My wife's parents agreed to care for her at their home, some 850 miles away from our farm. Upon our return, we went to retrieve her and were interested to see how she would react to her homecoming. Mink showed no unusual behaviours until our car was about ten miles from the farm, when she seemed to become alert and mildly excited. Arriving in the farm yard, she jumped from the car, made one circuit around the yard, stopping to sniff in the usual places, then went to her usual corner of the house, under the hollyhocks, layed herself down, and went to sleep! I realized that, at that moment, the past year had never happened for Mink. She was home, in her familiar surroundings, and all was well with her world.

The word "remember" itself provides significant insight if we insert a hyphen, to write it as re-member. The members of a body are its parts; arms, legs, head, etc. So to re-member someone means literally to reassemble the person we are thinking of, at least in our mind's eye. A very beautiful and profound illustration of this is found in the vision of the prophet Ezekiel, wherein God shows him how He will re-member his chosen people Israel. In Chapter 37, verse 1 we read[22]

> *The hand of the Lord was upon me; and he brought me out by the Spirit of the Lord and set me in the middle of a valley; it was full of bones.*

and, in verse 3 and following;

> *He asked me, "Son of man, can these bones live?" I said, "O Sovereign Lord, you alone know." Then he said to me, "Prophesy to these bones*

> *and say to them, 'Dry bones, hear the word of the Lord!' This is what the Sovereign Lord says to these bones: I will make breath enter you, and you will come to life. Then you will know that I am the Lord.'"*
>
> *So I prophesied as I was commanded. And as I was prophesying, there was a noise, a rattling sound, and the bones came together, bone to bone. I looked, and tendons and flesh appeared on them and skin covered them, but there was no breath in them. Then he said to me, "Prophesy to the breath; prophesy, son of man, and say to it, 'This is what the Sovereign Lord says: Come from the four winds, O breath, and breathe into these slain, that they may live.'"*

Perhaps we can also read this as a foretelling of the resurrection that is to come for all who believe.

Indeed, those who take the Bible as their Word of Life are constantly reminded to remember, remember, always remember. Remember the Ten Commandments, remember the Law, remember the Great Commandment and the Lord's Prayer. Not only are we called to remember, but we desire also to be remembered by God. We read in Luke[23] that the repentant thief on the cross called out to Jesus "Jesus, remember me when you come into your kingdom". He did not ask to be delivered from death, or from suffering, but only to be remembered. In the same vein, the Orthodox churches end a burial service with the prayer "Eternal Memory" which is a supplication that God will always remember the departed one; that he or she will always remain in God's memory. To be re-membered by

God is to have one's name written in the Book of Life, to be alive forever.

There is an act of remembrance that is of great importance for Christians of all denominations. That is the sacrament of Holy Communion, as instituted by Jesus with his disciples just before his crucifixion. The words of institution differ only slightly in the synoptic Gospels, and again in Chapter 11 of I Corinthians where Paul teaches that Jesus broke the bread, gave thanks, and gave it to them to eat, saying (paraphrased)

> *"This is my body, which is for you; whenever you do this you do it to remember me".*

Likewise he took the cup and gave it to them, saying

> *"This cup is the new covenant in my blood; do this, whenever you drink it, to remember me".*

I believe that the words "to remember" here mean, literally, "to re-member", in that, when we gather in one spirit to partake at the communion table or rail, we really do reconstitute the body of Jesus. At that moment he again has eyes and ears, our eyes and ears, to see and to hear the cries of the needy; also arms, legs, head, and trunk; our arms, legs, head, and trunk; to do what is his will to be done in this world.

When, however, there is dissension, jealousy, or anger between us, as happens all too often in the church, then the words and act of communion are of no good effect. Then, instead of re-membering Jesus, we dismember Him.

Chapter 7
FINAL FRAGMENTS

7.1 The Compassionate Timekeeper

Human life on Earth is not a game, but we could perhaps think of it in those terms for a moment. In most athletic games there is a timekeeper. The game has a beginning, when the game-clock reads zero, and an end, when the time-remaining clock reads zero. The timekeeper has, really, quite a lot of power to influence the play of the game. He or she can start or stop the clock, call the on-field officials over for a consultation, add more time to that remaining on the clock, and sound the final buzzer to end the game. Most timekeepers are strict in their adherence to the rules and procedures of the game; they do not grant favors to one team or the other, they do not forgive infractions, and they do not love one team or one player to the exclusion of another. Were they to do so, the excluded team could rightly protest "not fair". When the game is over, the timekeeper goes home, back to his or her real life. The game is just a game, and timekeeping is just a job.

Life on Earth had a beginning, and it will have an end. We do not see a game-clock or a time-remaining clock, but we believe that there is a timekeeper, namely God, the Alpha and the Omega. He, likewise, has a lot of power to influence what happens on Earth. He was here at the beginning and he will be here at the end, at the final time of his choosing. In contrast to the timekeeper of the previous paragraph, he does not often step in to enforce the rules. He has made known his perfect Law, and people are free to follow it or ignore it. There are consequences, though, for those who choose to live out their lives in disobedience. Also in contrast, God does forgive infractions for those who are truly sorry, and he loves all the players. He has stepped in, at halftime as it were, to make known his great love for this world and its people[24],

> But when the time had fully come, God sent his Son, born of a woman, born under law, to redeem those under law, that we might receive the full rights of sons.

Not many received him then. But the compassionate timekeeper has remained, he has not left us and has promised that he will not ever abandon us and "go home". His real life is here, timekeeping is not just a job.

Meanwhile, he has shown that he has authority over time, decay, and death. Of those who had been sick, some were healed; of those who had been blind, some were made to see; the dead have been raised. All these miracles involved resetting the clock for those who came to him in their need. He saw their need and had compassion for their suffering. In doing so the "clock" that was reset was not an ordinary timepiece but was, rather, the *entropic clock* for the individual. The ravages of disease and death were taken away as though they had never happened; life and health were restored to those of an earlier time, while ordinary clock and calendar time were not altered.

7.2 Light and Darkness, and Time

If you ask people "What is the opposite of light?" they might answer "darkness". This answer may be correct if we use the word "light" as a modifier or as a verb. It is *not* correct if we use it as a noun, meaning the physical phenomenon itself. There, darkness is the absence of light, not the opposite. It has no power to oppose or to overcome light; it is only the default state. All one can do to oppose light is to block it out, and, as Leonard Cohen wrote in his lyric[25]

> Ring the bells that still can ring
> Forget your perfect offering
> There is a crack in everything
> That's how the light gets in.

Brokenness is necessary, for the light to penetrate.

We understand that there are forces of darkness afoot in the world; those who would like to seal all the cracks and have no visibility for their motives and maneuvers. The means seem to be at hand. The idolatry of modernity – the preoccupation of all waking hours with entertainments and pastimes – is <u>designed</u> to leave no crack for light to come in. Yet there are cracks, for, as we read in John 1:5 "The light shines in the dark, and the darkness has never quenched it"[26].

So, getting back to the question, what is the opposite of light? Or, put another way, what is the opponent of light, what can neutralize or destroy it? I would like to suggest that it is time – entropic time. We saw in Chapter 2 that for objects travelling at the speed of light, namely photons, light itself, there is no elapsed time, as measured by conventional clocks. However, light is *diminished* by entropic time. The Penzias-Wilson microwave radiation, the residue of the intense light-energy at Big-Bang time, is greatly diminished in its intensity due to the entropic expansion of space. It corresponds to black-body radiation coming from an object (a black body) that is at the temperature of only 2.7 degrees Kelvin above absolute zero, or about -270 degrees Celsius. Contrast this with the radiation believed to be filling the universe immediately following the Big Bang event, which is believed to correspond to a temperature of about 10 billion degrees Kelvin at the time when nuclei began to form, about 1 second after Big Bang.

If entropic time is the enemy of light, then we might think that the war will end when time is no more, when Omega comes. We don't know what will become of light then, but in Revelation 22:5 we read "There will be no more night. They will not need the light of a lamp or the light of the sun, for the Lord God will give them light."[27]

7.3 Some Poems about Time, etc.

"The Gila Lion, or, an Argument against the Theory of Evolution"[28]

The deer came up from the creek this morning,
In careful steps, each hoof precise
In the frost-rimmed print of the one before.
Does and fawns, their heads and ears alert,
They came up this morning to my meadow,
They do not fear me, nor any man.

The lion came down the mountain last night,
The lion came down in darkness,
Came down on soft pads, eyepools of starlight,
Came down in hunger, days since his last kill.
He is hunger, waiting,
Waiting above my meadow now.

Perhaps the deer know this; it is their nature
To know such things. Perhaps they know
That he will take just one
In the flash of tawny fur and blood,
And the others, moving off a little way,
Will resume their browsing, while the lion feeds.

I see all this, and yet I do not
See how fear and evil came,
To my neighbor's heart, and mine.

"Ghost Towns of New Mexico"

You see them everywhere,
On eastern plains and beneath the mountain peaks,
Collapsing heaps of sunbleached boards,
Where folks once lived and worked and laughed,
And loved, in soft lamplight when work was done,
And often wept, when hopes and dreams were gone.

None left now, the young are off to brighter lights
Of cities where the laughter never stops
And the tablets glow and the young
Know there no fear of dark and lonely nights,
While the old, out beneath their weathered crosses
Are forgotten, but they weep no more.

Though they sometimes wander in, it seems,
To stand above, on hilltops, in their dusty clothes,
To watch the cities burn, to pray, and then
To turn and slowly wander out again,
Back to their timeless watch for the dawn
Of the light of the world to come.

"Traces"

He was surprised, he said, at grief, at its manner
Of overtaking, from without, as a wave crashes
On the unsuspecting shore. Forty years with her
Had left him unprepared, inconsolable with ashes
Finally left among the roses that she loved.

But still she came to him, it seemed, in moments
When a certain string was touched, in memory
Unbidden, nearly there, slightly present
Among the traces of their life, taking briefly
Breath from him then gone again.

Exhaustion gave him no release, hard labor
In the meadow, in heat burning from the rock
On narrow valley sides, heaving bales aboard
The wagon, evenings trailing the silent sheep
To their fold beside the murmuring stream.

At any moment there might be a hint, a faint
Movement in the corner of the eye, a sigh
Of breeze that brought the thought of her again
To image in his mind, her calling to come nigh,
To come to her, to end, to bring to her their life.

And in the starry nights, when sleep did come
At last, his arms might move without his will
Embrace beside him empty space where once
Was warmth and welcoming, her scent still
Lingering on the bedding of his mind.

But moments passed, and he would turn
His face towards his wall and sense
The world close in again, and wait for dawn
Another day to try once more to balance,
Massively, against each fleeting trace.

"Bringing in the Hay"

Through white and silent winter night an old man dreams
Dreams of green and gleaming days of early summer
    When hay stood tall in meadows and men and boys with horses did
    What we had always done, the harvest of the hay in wagons fragrant
        Loaded tall with stacks heaved high and joy
        was won through laughter
        Streaked with thirst and salty sweat and
        pain.

There's old Jake driving his good team Nell and Bess
Standing on the wagon tongue, on his stubbled face that rascal grin
    While his load sways and almost tips as they careen around the field
    Boys shout and skip for mirth at doing this great work together, at being
        In the procession of the wagons bringing in the winter's hay
        Through halls of sun drenched dusty summer gold.

But at the end there is a great and closing door
One runs to let the wagons through, and with the click
    Of closing latch comes silence, empty corridors of softly glowing
    Varnished wood in afternoon of autumn light, and down and down

A silent stair with burnished balustrades, one hears only faint
And distant voices, and finally there the darkening farmyard

Horses and wagons gone, but where now a Figure comes
To find the one who has fallen out of time.

"Why Lambs Die"

The wind, the wind drives ice
Needles through my coat where there
The lamb trembles as I trudge
Away from the fragrant barn.

Only hours ago the lamb was born
To warm breath of beasts, answering
The ewe's nickering call
With its own small voice.

But then an icy finger touched
And beckoned, chilling still damp
Flanks and stilling
That small lamb voice.

I know what next - beside my fire the lamb
Will warm and dry, will even weakly suck a rubber teat,
But won't answer my call and soon will turn its face away and die
Of ambiguity.

"Heartwood"

Some trees - the willow,
Pine, and fir, perhaps the yew?
Writhe inside as they grow old,
Binding death within the core
Of waiting heartwood, dense, and hard,
And deep, within the massive bole.

Portent of death, and yet,
Of dusky beauty blazing forth
In sawn plank, in skeins and whorls
Of ancient ochre shot through young and yellow bands,
Like heartwood of our life, my wife of many years,
Of beauty found when time shall wear the sapwood down.

7.4 The Invitation

Some time ago I had a dream. "So what", you say, "We all dream every night". But this dream seemed different. It was one of those dreams that, unlike most dreams, aren't forgotten immediately upon awakening, but stay with us for a few minutes or hours. "Big deal", you say, "We've all had at least a few dreams like that". But this was a vivid and disturbing dream that would not be put aside, and it continued to nag me for days. It seemed that it might contain a message of some sort, one that demanded understanding. And so I spent some time during the next several days thinking about it, wondering why I had such a dream and what, if anything, it might mean.

The Dream

In my dream, I was driving somewhere, on some routine trip or other, when my route took me near to the rural home of a relative. Exactly which relative I don't remember, but I decided to stop for a visit. It was only to be a brief stop on the way to somewhere else. When I arrived I found, to my astonishment, that there was a party going on, a great gathering of people, something like a grand reunion of all my extended family and many others whom I had known. Among the multitude I saw my sister who had died nine years previously; I saw a favorite cousin whom I hadn't seen in 40 years.

Several people greeted me, but seemed surprised, as though they wondered why I was just arriving and appeared to be confused. About then I realized that I hadn't been informed about the party, and that I wasn't suitably attired, coming as I was in my everyday clothes. I also realized that I couldn't stay, as I hadn't been invited and had other places to go that day. So I spoke to a few people, who seemed puzzled as to why I hadn't been invited, and then I left in great sadness, wondering whether someone had failed to mail my invitation, or if my name was intentionally omitted from the guest list.

In thinking about this dream, I had to wonder if it might be a premonition of something to come. There came to me the thought that each of us has a pocket full of invitations, one addressed to each member of our immediate and extended family, also one for each friend and neighbor, past, present, and future. Our job is simply to deliver the invitations. Perhaps the invitation might read something like

> *You are invited to a party. The party is called* **The Gathering at the End of Time**. *All that is needed will be provided. Dress is not optional, but a suitable garment will be provided. No gifts, please, only your presence is desired*
> *R.S.V.P. to "The Lamb."*

If you, dear reader, having come this far in this little book, have never received such an invitation, then this one is for you. Please respond soon - time slips away. If you don't have words to instruct your heart in this matter, perhaps the following verse from a familiar old hymn might be helpful.

> *Just as I am, without one plea*
> *But that thy blood was shed for me*
> *And as thou biddst me come to thee*
> *O Lamb of God, I come*
> *I come*

REFERENCES

1 Dylan Thomas, from THE POEMS OF DYLAN THOMAS, copyright ©1945 by The Trustees for the Copyrights of Dylan Thomas. Reprinted by permission of New Directions Publishing Corp.

2 Anne Dillard, published in Atlantic Monthly, Vol. 242, October 1978

3 Berry, Wendell I, 1993. Copyright © 1998 by Wendell Berry, from *A Timbered Choir*, reprinted by permission of Counterpoint Press.

4 Gerard Manley Hopkins, from Poems of Gerard Manley Hopkins (Oxford, 1948)

5 Newton, Sir Isaac, Philosophiae Naturalis Principia Mathematica, 1689

6 Newton, *ibid*

7 Brian Greene, *The Fabric of the Cosmos: Space, Time, and the Texture of Reality* (Knopf, NY, 2003)

8 William Butler Yeats, The Second Coming, from Collected Poems by W. B. Yeats (McMillan, 1951)

9 Schaeffer, Francis, How Should We Then Live? (Crossway, Wheaton Ill 1976)

10 Harari, Yuval Noah, Sapiens: A Brief History of Humankind (McClelland & Stewart, 2014)

11 Harari, Yuval Noah, Homo Deus: A Brief History of Tomorrow (McClelland & Stewart, 2015)

12 Penrose, Roger; The Road to Reality, A Complete Guide to the Laws of the Universe, pp. 729-730. (Vintage, 2005)

13 Kenneth C Steven, Prayer, from Iona, Poems (St Andrew Press, Edinburgh, 2000)

14 Genesis 1:1-5 (Taken from the HOLY BIBLE: NEW INTERNATIONAL VERSION, copyright 1978 by the New York International Bible Society, used by permission of Zondervan Bible Publishers).

15 Romans 8: 19-21 (Taken from the HOLY BIBLE: NEW INTERNATIONAL VERSION, copyright 1978 by the New York International Bible Society, used by permission of Zondervan Bible Publishers).

16 Dylan Thomas, Do Not Go Gentle into that Good Night, from Miscellany One (J. M. Dent & Sons, London, 1963)

17 C. S. Lewis, The Great Divorce, p. 66 (Geoffrey Biles, UK, 1945)

18 Matthew 7:23 (Taken from the HOLY BIBLE: NEW INTERNATIONAL (Taken from the HOLY BIBLE: NEW INTERNATIONAL VERSION, copyright 1978 by the New York International Bible Society, used by permission of Zondervan Bible Publishers).

19 Revelation 21:1-4 (Taken from the HOLY BIBLE: NEW INTERNATIONAL VERSION, copyright 1978 by the New York International Bible Society, used by permission of Zondervan Bible Publishers).

20 Luke 23:43 (Taken from the HOLY BIBLE: NEW INTERNATIONAL VERSION, copyright 1978 by the New York International Bible Society, used by permission of Zondervan Bible Publishers).

21 H. F. Lyte, Abide with Me (1847)

22 Ezekiel 37:1-9 (Taken from the HOLY BIBLE: NEW INTERNATIONAL VERSION, copyright 1978 by the New York International Bible Society, used by permission of Zondervan Bible Publishers).

23 Luke 23:42 (Taken from the HOLY BIBLE: NEW INTERNATIONAL VERSION, copyright 1978 by the New York International Bible Society, used by permission of Zondervan Bible Publishers).

24 Galatians 4:4 (Taken from the HOLY BIBLE: NEW INTERNATIONAL VERSION, copyright 1978 by the New York International Bible Society, used by permission of Zondervan Bible Publishers).

25 Leonard Cohen, There is a Crack in Everything, from Leonard Cohen, Selected Poems, 1956-1968 (McClelland & Stewart, 1968)

26 John 1:5 (Taken from THE NEW ENGLISH BIBLE, copyright 1961 by The Delegates of the Oxford University Press and The Syndics of the Cambridge University Press)

27 Revelation 22:5 (Taken from the HOLY BIBLE: NEW INTERNATIONAL VERSION, copyright 1978 by the New York International Bible Society, used by permission of Zondervan Bible Publishers).

28 Ray E. Rink, previously published in the Mimbres Messenger, February 2016

Acknowledgements

I must acknowledge a lifetime of intellectual stimulation and spiritual guidance received from many friends, colleagues, pastors, and teachers, too many to mention by name. I will mention one name, the late James A. Keller, formerly Profesor of Philosophy at Wofford College in Spartanburg, SC. Jim was a lifelong friend and mentor and had a great influence on my life during early years at MIT. In recent decades our philosophical paths diverged, his towards process theism while mine remained with evangelical Christianity. Nevertheless, he continued to encourage me to undertake the writing of this book. I would also acknowledge the support and encouragement of the good people at Westbow Press, who were always willing to help and advise, even though I wasn't always willing to listen. The consequent defects of style are mine. Last, but not least, I must acknowledge a lifetime of love and support from my wife, Beverly.

Cover Artist William F. Quast

Bill Quast studied art and music at Concordia Teachers' College, now Concordia University of Chicago. He has had one-man exhibits at his alma mater, at the the offices of the Canadian Consulate General in Chicago, at Valparaiso University in Indiana, and at the Multi-Cultural Heritage Centre in Stony Plain, Alberta. He is a retired teacher, having taught art, music, and religious studies at Concordia University, Edmonton.

About the Author

The author is a retired engineer and lives with his wife on a small farm in north-central Alberta. They also spend some winter months at a cabin in the Gila Mountains of southwestern New Mexico. He received post-secondary education at MIT (BS, 1962) and at University of New Mexico (PhD, 1967). He has been employed by Sandia Corporation, Albuquerque NM, by Oregon State University and by the University of Alberta, and most recently as an engineering consultant to the oil-sands industry in northern Alberta. The author enjoys reading, writing, hiking with his wife and dogs, and flying airplanes.

www.ingramcontent.com/pod-product-compliance
Lightning Source LLC
Chambersburg PA
CBHW031924240526
45464CB00022B/789